WINNER

Innovative Ideas for a Better Tomorrow

For Startups, Businesses and Applications

I0395940

SHAIK KHAJA SHAHABUDDIN

INDIA · SINGAPORE · MALAYSIA

Notion Press

No.8, 3rd Cross Street,
CIT Colony, Mylapore,
Chennai, Tamil Nadu – 600004

First Published by Notion Press 2021
Copyright © Shaik Khaja Shahabuddin 2021
All Rights Reserved.

ISBN 978-1-63745-388-9

This book has been published with all efforts taken to make the material error-free after the consent of the author. However, the author and the publisher do not assume and hereby disclaim any liability to any party for any loss, damage, or disruption caused by errors or omissions, whether such errors or omissions result from negligence, accident, or any other cause.

While every effort has been made to avoid any mistake or omission, this publication is being sold on the condition and understanding that neither the author nor the publishers or printers would be liable in any manner to any person by reason of any mistake or omission in this publication or for any action taken or omitted to be taken or advice rendered or accepted on the basis of this work. For any defect in printing or binding the publishers will be liable only to replace the defective copy by another copy of this work then available.

CONTENTS

Author's Note   7

Chapter 1   Isleep, Smart Sleep, Home Office, Sleepulator   13

Chapter 2   App for Housewives for Food Preparation, Distribution, Identity Hidden and with Ratings   24

Chapter 3   Washing Machine, Commercial Vehicle Visiting near to Our Homes on Demand or on a Regular Daily Routine   30

Chapter 4   Salon on Wheels, Vanity Van Concept   35

Chapter 5   Health Consciousness and Awareness for Morning Walkers Including Senior Citizens   39

Chapter 6   Roaming Concept (Volunteers, Apps, Camera), World Tour, Virtual Tour for all Financial Backgrounds, Economically Weak/Strong Individuals/Groups   44

Chapter 7   W/C Transformation for Better Time Quality and Enjoyable   51

Chapter 8   100 Cities 100 Professionals   55

Chapter 9   Apps for Anti/E-Commerce Giants   63

Chapter 10	Robots Creation, Roaming Purpose	67
Chapter 11	Video Camera Cum Main Door	72
Chapter 12	Transportation without Traffic Jams	75
Chapter 13	Religious Places Location	78
Chapter 14	Public Representative Accountability and Betterment of Social Life	84
Chapter 15	Mobile Toilet with A/C and Ventilation in a Commercial Vehicle	90
Chapter 16	Household Expenses Management Team	96
Chapter 17	Women/Girls Protection through an App	102
Chapter 18	Modern Shower System with an Injection System	107
Chapter 19	Pedalling Cycle for Power Generation	111
Chapter 20	Apps for Endorsing Local-Made Products	113
Chapter 21	Local Tourism Holidaying	115
Chapter 22	Hotel/Restaurant all Over the City	119
Chapter 23	Career Guidance/Investment Guidance/Loss Prevention/Advisory Committee gives Service at Some Charge	122
Chapter 24	Bed Modified which can Accommodate A/C, Heater, Fan as per the Weather Conditions for Getting Comfort Air Temperature	125
Chapter 25	Protection of Indian Citizens around the World	130

Chapter 26	Protection Suits for Two-Wheeler Riders and Passengers, Non-Ac Car Drivers and Passengers	136
Chapter 27	Flying Vehicle Rapiport	139
Chapter 28	Movies 4D Concept with Screens	143
Chapter 29	Water Bottles Designed for Displaying	148
Chapter 30	Innovation of Special Goggles	152
Chapter 31	Innovation of Bed Capsule Type	155
Chapter 32	Innovative/Informative Hands-Free Suitcase	160
Chapter 33	Innovative Pen with Cartridge	163
Chapter 34	Innovative Shoes/Chappals which Display the Number of Steps Taken with a Reset Option and Display the Weight	165
Chapter 35	Innovative Toilet Seat	167
Chapter 36	Ladies Special Vanity Van	168
Chapter 37	Ladies Special get Together in Every Area Individually	169
Chapter 38	Vending Machines	172
Chapter 39	National Private Teams	175
Chapter 40	Innovative Goggles and Headphones in One Piece	180
Chapter 41	Winning Over the Death	185
Chapter 42	Becoming Invisible or Camouflaging Concept and Protecting Our Loved Ones/Borders/Business Places/ Homes/Vips/Homeland	196

AUTHOR'S NOTE

INTRODUCTION, INSPIRATION AND PURPOSE

To all the readers, a warm welcome for taking out time to read my book.

I being a double-degree holder had never found happiness in doing any kind of regular jobs, even though I completed my projects successfully and got good salary/perks (my first job abroad in the year 2000 got me a salary of 1000 US $/month).

By the time I realised that I had to do something in my life which gives me satisfaction or happiness and allows me to follow my passion to innovate/create/improvise and rule the world through my ideas, 20 years had passed from the day I got out of college or stepped into the job world.

Today, I am in my early 40s and realised that I have got a special power from the experiences/failures/success stories/the way I handled my problems/the travel across the globe/coming back to my homeland/working in companies abroad/interacting with the people around the world/reading people/understanding different cultures.

The world has changed so much from 2000 until today. I had a vision and passion for innovations or improvisations right from the day I stepped into the job world in early 2000. I was busy fulfiling someone else's dream and people followed their passion and launched in the market so many products that once I thought were useful and the idea had stuck in my mind as well, for e.g., the delivery apps/social networking sites/the innovations/online shopping apps/e-commerce apps/ education apps.

So better late than never, the main purpose of penning down my ideas is to break the barrier of criticism/self-suppression, take everything positively and give the world the future technology.

My book will be in detail about new technologies, innovations and improvisations, making lives better, creating jobs, helping each and every individual realise his/her potential, creating apps for all, not to leave behind anybody, and guiding them to the path of success.

Each innovation and app will bring out thousands of jobs, make lives better, make everybody responsible citizens, make one realise his or her potential, and make everyone accountable for their responsibilities in the society, and I am proud to say that the whole world will follow the footsteps of my great country, INDIA.

I have not left out even the underprivileged; I have made innovations and apps useful for kids/elders/ working women/single mothers/senior citizens/ housewives/working class/business class/students/ professionals/engineers/doctors/lawyers/CAs/sports players/cleaners/daily wage workers/labourers/idle roaming people/public representatives; each and everybody will be benefitted with the help of my innovations/apps, creating millions of jobs, making our country strong, and helping in eradicating poverty, making lives better, eliminating unemployment and making everybody self-dependent financially/physically/emotionally with social responsibility.

The main purpose of my innovations and apps will be to bring all the comforts/luxuries/services/technologies to the doorstep of a common man (lower class, middle class, upper-middle-class, and economically weaker sections) at a very affordable and cheap price.

The inspiration was divine and I am thankful to my CREATOR ALLAH SWT for filling the empty/confused brain to grasp, understand, and motivate myself to write and spread out the knowledge in the form of this book.

The brain has the capacity/capability to travel in the universe, go to the past, travel in the future, bring the future technology and apply it in the present to help the mankind/improvise lives/eradicate poverty/eradicate unemployment/eradicate diseases/be prepared for the disasters or to avoid them/be prepared for the pandemics and epidemics/make a global village where everyone is in the process of knowing, helping and supporting each other.

Finally, I would like to inform that my duty will be fulfiled when all the readers realise their potential and feel no less than anybody and bring out the best in any form like ideas/courtesy/innovations/improvisations/behaviour/willingness to help humanity/character/nature/social relations/family relations/helping future generations and making the world a better place in terms of faith/offerings/character/social life/behaviour/dealings.

The main purpose is to make all my ideas/innovations/apps a reality and I will be working towards it and help humanity to be successful in every aspect.

Companies/individuals/corporates/intellectuals/visionary, passionate persons/MNCs/educated, uneducated, underprivileged/celebrities/famous personalities/experienced/amateurs/skilled, unskilled/talented/idle, roaming persons/homemakers/singles/widowed/married/senior citizens/cleaners/daily labourers each and everybody can come forward and help build a GREAT NATION by contributing and making this a reality.

Finally, I am thankful to all the people in my life who have accepted or rejected me, tried or tested me, utilised or neglected me, boasted or criticised me, loved or hated me, understood or misunderstood me, carried or dumped me, given me or snatched away from me, helped or cheated me, favoured or stranded me, and used me or thrown me, for making me stronger and helping me realise my potential and they taught me to never give up in any situation/crisis.

My life was like a roller-coaster ride until now and even so will be in the future with ups and downs; but every day of my life, I will grow stronger and stronger to fight the problems/unfavourable conditions/failures/defeats/criticisms/hatred and still be thankful to the ALMIGHTY ALLAH SWT for giving me a chance to be born as a human being which is HIS best creation; my aim is to guide the humanity and I will never give up;

I keep working on it through my ideas, deeds, motives, plans, innovations, passion and vision.

HAVE A SUCCESSFUL LIFE, NEVER LOOSE HOPE, HELP AND SUCCESS IS VERY NEAR

SHAIK KHAJA SHAHABUDDIN (A GREAT LEARNER)

I dedicate this book to my family members who have been supportive in all phases of my life, accepted me in every situation, never left me alone, and been an inspiration through their unconditional love, unambiguous service/care/respect.

ANY INDIVIDUAL/COMPANY THAT WANTS TO USE MY IDEAS AND DEVELOP APPS/INNOVATIONS CAN CONTACT ME AND GET THE COPYRIGHT.

Chapter 1

ISLEEP, SMART SLEEP, HOME OFFICE, SLEEPULATOR

Bed capsule type with all the features as that of a smartphone, home theatre 4D concept, screens all around look like a shell, door opening with a retina scan and fingerprint.

Hydraulic or pneumatic operated simulator with all-around screens inside, orthopaedic bed concept, climate control, with heating and cooling mechanisms for comfort, air, temperature and destination place temperatures for experiencing the real effects while live video viewing.

Capable of making tests for bp/sugar/pulse and keeping a track record of all the data, with telescope fitted on the roof and transmitting the images to the screens of bed, all four views transmitting onto the four screens inside, food and drinks door and display both inside/outside, motion sensor.

Connected with the roaming concept, live recording and broadcasting directly with simulation effects, the power from a DC battery source inbuilt but charging from the household domestic power supply, themes of sleeping and climatic effects like the ambience of any desired place, travelling and night view from telescope or else any other country, sunrise and sunset at any time.

All the necessary software for con calls, video calls, creating short cartoon movies, creating heavens for self, designing buildings and flying palaces, designing with all interiors and mountains and rivers, designing and flowing as per requirement, short films using all the characters available including music creation, putting out morning thought through video or a comedy bit available for everybody to comment as that of same like posting videos, roaming concept accessibility.

Office work, video conferencing, viewing the site status for huge constructions through drones, attending meetings, teleconference, recording all health data for all individuals for health records and analysis (couples, singles, elders and kids).

Lost loved ones' meetings in their heaven and interacting with them through the manufacturer's latest software.

Air, positive pressure and stopping poisonous gases and earthquake resistant and protection and along with hooter and lifesaving necessities, oxygen cylinder connected to supply comfort oxygen levels and shut off valve if poisonous gases enter the inside and disaster safety features.

Recording self-videos and motivational speeches and self-experience videos on how to tackle problems in life and whatever is happening in life or record the whole day's experience and explain lessons learnt and mistakes and good deeds done on that particular day.

Heaven visit installed by the manufacturer or creating own heaven by making and beautification of imagination places, castles, rivers, mountains, with loaded software's colour schemes, and gravityless environment all in 4D.

Heaven visiting and exploring loaded by the manufacturer, 4D experience, 4D movies, mythological movies related to all Holy books and being a part of the detailed life of those greatest messengers within the movies, and learning human values, we can be a part of all the stories of our holy books and experience in these 4D movies with real effects loaded by the manufacturer.

Softwares for introspection by imagining in our own grave and recollecting and remembering all the good deeds and bad deeds will be weighed and then we will be allowed or denied entry into heaven.

Movies, short films, using self-character or the characters loaded by the manufacturer and putting up for viewing for others and can win the best short film award or else can get the copyright for making a real 4D film.

BHARAT VERSH

SOFTWARES INSTALLED WILL BE TERMED 'B-VERS' MEANS 'BHARATH VERSIONS' OF ALL THE ABOVE MENTIONED PROGRAMMES. CREATED, DEVELOPED AND INSTALLED BY INDIAN INTELLECTUALS AND PASSIONATE INNOVATIVE KEY PLAYERS AND VISIONARIES.

INTRODUCTION/PURPOSE

A bed with screens all around inside, more like a tube and oval-shaped enclosure with hydraulic or pneumatic jacks and simulation technology with all the available software old and latest, 4D theatre and with air-conditioning and heater, ventilation pressurised to keep positive pressure, gas-proof for stopping dangerous gases to enter inside, oxygen supply with the cylinder placed away from or on a rooftop and door and digital display outside with a camera, break-proof material, door and CCTV for surveillance with a facility for relieving or passing urine with pipe and water washing vacuum or suction type for elders and sleeping persons unable or not to disturb themselves by walking to the washroom.

A 7'X7' round or oval-shaped bed dome type and its features will have all the latest software's old and new and developed, AC/DC powered and voice-activated fingerprint lock and retina scan for unlocking along with the keypad lock and pin.

DETAILS/DESCRIPTION

- A bed with hydraulic and pneumatic jacks and simulation technology.

- Telescope viewing, orthopaedic bed and imitation of movement yaw, pitch, banking, diving and all roaming videos will be modified as per the real capturing effects, transmitted/viewed/experienced.

- Softwares preloaded for making short movies with all the characters and music creation and using self-character and all movie and

TV characters and political or any famous personalities.

- Softwares for car designing, house designing, heaven designing and unimaginable things and universe creation and videos for viewing from manufacturer creations loaded by a company.

- Heaven visiting, embracing and appreciating the unimaginable world and watching with all the 4D effects.

- Sleeping while the themes are running as screen savers and those will be hundreds in number for ambience or display of any part of the world and any effects in 4D like airflow, snow, fragrance, water spray, rain effect, thunderstorm and lightning, etc.

- Sleeping while themes are running in screens or virtual travel to any part of the world and means of transports changing as per requirement and sound effects or visiting space, all starting from home to reaching any country by flight, train, bus and luxury cars and will play till morning and even travel in water by ship.

- Preloaded videos for rent or purchase for journeys to any part of the world by any means and enjoying the world tour at a very less cost comparatively by real travel.

- Oxygen levels and CO_2 levels checking and supply of pure air and checking for dangerous gases in case of mass gas leaks and will not allow inside the den and in case of gas leak, a hooter and survival technology in disaster and case of any natural calamity.

- In case of an earthquake, the body will resist any damage from outside as well as inside and protect the occupants.

- Simulation of any travel by any means as a traveller or self-driving mode.

- Preloaded series of mythological 4D films and experiencing the real effects of being a part of the events and the story.

- Preloaded videos of historical events and experiences by being a part of those events related to all history characters.

- Visiting our loved ones in their heaven and interacting with them and going on a tour with them to visit their special places.

- Visiting our own heaven and experiencing the unseen and unimaginable scenic and natural beauties and the manufacturers will keep on developing and releasing newer versions of heaven for an experience of latest effects in 4D.

- The roaming videos will be developed and put out for rent along with the effect of the simulation for flying, running and travelling and even the odours and airspeed and will be calculated and even with temperature and other climate effects as in the real one while visiting for world-tour videos.

- All the roaming videos can be seen from the four-side camera and the real effects can be experienced in a 4D movie, which includes simulation.

- One can make films with subtitles, sound effects and music, put out in public domain, get a copyright for those content, win awards and can sell the movie for real film making in 4D mode with real characters selected by the owner.

- We can make our own FUN videos, make use of all the characters and put in the public domain as a status.

- The short-film concept allows any individual to put up his or her ideas and if selected, one can get the copyright of the content, sell it to those who want to pay and make a real film and become a potential storywriter and filmmaker.

- By making use of all the preloaded software, any individual can become a director, storywriter, music composer, building designer, car designer or any game designer and can utilise his or her talent, with no restrictions whatsoever.

- Sleeping in any ambience in any part of the world and experiencing the effect of water, air, speed, snow, and changing seasons in 4D and even travelling, running, walking and flying and experiencing the effects by jerks and speed breakers and bumps and uphill and downhill effects like we are really travelling that sort of simulation will be experienced.

- The tablet wireless, INFOTAINMENT GADGET will be used as well for all operations, controlling, playing and using all features.

- Even the rented robots used in roaming can be controlled with the tablet.
- Real-time roaming videos will be broadcasted directly.
- The future 4D cinema concept.
- The concept of relieving and washing is pre-fitted WITH-IN.
- The hydraulic/pneumatic jacks, simulator technology and telescope for transmitting/watching and the drain connections, and the compressor kept at the roof.
- The work from home concept is full-fledged for meetings, con-calls, conferences and zoom meetings.
- Ordering of food and receiving a special door from the home kitchen with the display on the outside screen and a camera to cover all places/views inside the house.
- The journey around the world with effects will be preloaded and can be rented with real 4D effects; preloaded roaming, live roaming or robot roaming will be available.
- By sitting and sleeping, we can travel around the world, without strain and finish all the office work, with a multitasking feature of the system and multitasking split-screen if necessary.
- Earning while making short films, composing music, building designing, interiors designing, making cars, bikes and furniture, and dresses

and fashion designing; with the software, we can create anything in our imagination.

- Creating our own heaven with unseen and unimaginable things and patterns and colours and animals, rivers, mountains, planets, universes, places, dresses, rivers, grass, sky, clouds, rivers flowing above us in the sky and all unimaginable contents.

- Preloaded heaven from a company and in that, we can visit, tour, travel and see never seen things and creations and visiting our own angels and deities and visiting them as per our beliefs and as per our faith.

- The air-conditioning and heating mechanisms, pressurised pure air supply, stopping by sensing dangerous gases, leaks and gas-leak attacks or any disaster or natural calamity all the features will be there for our protection.

- People with lack of sleep will be relieved of their deficiency, because of the ambience features, roaming concept and 4D effects.

- In future, all the ads, films and news broadcasting will be done in the 4D concept.

- By utilising the available software, anybody can follow their passion, can copyright the product or content or creation, for e.g., for fashion designing, short/long film making, serials making, web series making, ad maker, car designer, bike designer, flying vehicle designer, architect, interior designer, heaven creator, furniture creation, and funny videos creation.

- Four-sides screen combined to watch 4D movies or else multitasking screens for office as well as music listening and movies watching, news watching and designing and multitasks at the same time.

- For getting the real effects of the preloaded videos or live videos cartridge for smells and odour, and temperature sensing and simulating and exactly replicating.

- The facility of replicating any ambience and any weather conditions or else any part of the night and day and any travelling and roaming around any part of the world at any given time helps in getting sound sleep and satisfaction, for e.g., if it is an afternoon and you are a late sleeper, but the ambience effect while you are waking up will be as that of sunrise for the feeling of freshness and or else the effect of the rainy day, snow day, Himalayas, beach, etc. can be purely felt inside the den.

- All the journeys, travels and ambience will be recorded and whatever we can imagine like we are in heaven, we can select the settings of ambience.

- The orthopaedic bed with different temperature settings for partners and respiration and special pillows and bedsheets like anti-bacterial and anti-fungal will be provided.

- The telescope will give live feed from the night or day even for watching the planets, the stars and the eclipses and live feed from space in 4D.

- The main purpose of this den bed and enclosure is to get a sound sleep as per the requirement of the user of different age groups and using their imaginations for the light settings as well as ambience and the aroma and their perception and their choice of ambience, and simultaneously becoming an entrepreneur and following passion by becoming a designer and exploring ambitions and enjoying life to the fullest, and experiencing the life and celebrating life and bringing the world to one's bed and doing the office work and monitoring one's health, enjoying work, feeling love, living, enjoying the movies in 4D and being in touch with the whole world, as well as present yourselves to the world.

Chapter 2

APP FOR HOUSEWIVES FOR FOOD PREPARATION, DISTRIBUTION, IDENTITY HIDDEN AND WITH RATINGS

INTRODUCTION AND PURPOSE OF THE APP

The main purpose of creating this app is to show the power of women who are homemakers and housewives, single mothers, what they have been cooking for family members and how they are underestimated for their hard work to cook good food with minimum expenses to make their family happy and healthy.

The purpose is to bring out the best talent and skill from inside of the women, whose work is neglected every day even after continuously dedicating their life just to keep their family happy and bind them together; they have been working 24×7×364 nonstop without even getting appreciation and remuneration and still, they are ignored and blamed for not giving enough time for the family and not doing anything special.

The household women have been cooking for years; when they are given a value and appreciation from the outsiders, how much they can earn doing their regular works, with the same amount of hard work and time! Normally, we appreciate the food cooked by others and are always unhappy with our own homely food, no matter how much hygienic and tasty it may be, which is a total disaster on our part in thanksgiving, respecting and appreciating the women of our household.

DETAILED EXPLANATION

- When a women or girl cooks food for the household, she knows exactly the quantities of ingredients, so nobody will calculate the food cost while preparing home food.

- So by regularly cooking for 5/6/7 family members, if extra food is cooked for 5 or 6 people like normally they do while some relatives or friends visit them in many festivals and occasions, they can get paid for that food by delivering to the people desperate for home food (who want to avoid restaurant food or roadside food and are not capable of cooking for whatever reason) and the price will also be dead cheap and affordable, and in return can cover for the food cost at home.

- The jobless single mothers and widows can be given accommodation around the country in each and every city and cover every area in that city so that they can cook and earn at the same time and support their family.

- On the other hand, the customers will get the hot homely hygienic food at home or office at a cheaper price comparatively with restaurants, which in return will help and even take care of the food expenses for food maker.

- Like, for e.g., the food prepared in the restaurant needs some calculation for deciding the selling price, which includes food cost, labour charges, rentals, salaries, power bill, taxes, wastages, and finally the profit margin and percentages of the partner(s) and/or owner; so automatically the price goes up for any delicacy or item prepared in a restaurant compared to that prepared at home.

- This concept will help both the customer and food maker; the customer will get hot food as per his choice at a convenient time at his required place and at the same time, the food cost will be covered for the one who is preparing and packing/sending.

- This concept and app will help those people and individuals who will never have to think about the cost or price of the product or item and can enjoy the delicacy like that of a restaurant/hotel/food chain and yet get healthy hygienic and home-cooked food and even for those people who are fed up of their own home food and would like to have home food but from other homes for a change.

- The concept and app will help those elderly people or senior citizens who are dependent upon the household women or maids for their meals and timely food and snacks, and the household women sometimes are busy with some personal works and are in a not good mood or may have some urgent appointments and may have to go out, but can't leave them without food, yet to fulfil their duties and may have to miss their doctor appointments and may further ruin their health.

- This concept and app will help those families in which the whole and sole cook like wife, sister, daughter-in-law, or daughter is sick and all of a sudden, kitchen stops food preparation and serving the household members for their meals, and then the only option left is to get some relative or extra help or ordering outside food, but in this app and concept, they can get food delivered to their home which is home-cooked food and even food for sick and doctor-recommended food and diet and still be carrying on with their own activities and taking care of the sick.

- This app and concept allow ordering desired food for breakfast, lunch, evening snacks and tiffin, and even dinner with the facility of preordering.

- Even the food maker will put out the menu one day in advance for the next day and the customers can pre-book and all the items will be displayed along with prices and quantities for serving no. of pax.

- The identities will be confidential and each user will be allotted a user name prescribed by the app developer and the rating for food maker upon likes will be given from 1 to 10.

- There will be a third-party food delivery agent involved who will deliver the food to the desired location upon the requested time.

- The amount will be prepaid including delivery charges.

- Ratings will be given by the customer to each cook and will be displayed in the public domain for the customer's ease of ordering.

- The food maker can display the items, which will be prepared as per the routine or weekly menu set by the cook.

- The packing material will be food-grade and can be preordered from the company.

- To protect the identities and avoid any stalking, only the delivery persons are authorised to deliver food at minimum delivery charges, which will have a range area or else an extra cost to any area.

- Likes and dislikes for the quality and quantity will be put out in the public domain to help improvisation.

- All the servings and quantities will be predetermined by the company including the weight of raw material used, for e.g., 250 g chicken curry or else 500 g rice or ½ kg of potato used for curry preparation; all the veg food makers will be given green or red for a mix of veg and non-veg and for fish as well including the display of halal as per the customer requirement.

- The app will have an option to deliver food as per the medical condition of patients and what is prescribed by the doctor, less spicy, less oil, no oil or only olive oil and quantities for all the items will be in weight only.

- Any quality issues will not be tolerated and if they have, they can get the money refunded.

- Fast foods like Chinese or pizza or burgers can also be ordered including cakes.

- Finally, benefits of home food compared to the dhaba/restaurant/roadside eateries or food chains are obvious, no need to worry about home food delivered at a time to your desired location, no need to carry tiffin boxes, Tupperware or hot cases to keep food hot and still get home food but from different home who care for their customers.

- The app can be downloaded which is safe for the individuals preparing food and the customers at the receiving end.

- If the ratings are good, then the party orders can also be taken and the delivery company will have a special packing and delivery procedure at an extra cost.

Chapter 3

WASHING MACHINE, COMMERCIAL VEHICLE VISITING NEAR TO OUR HOMES ON DEMAND OR ON A REGULAR DAILY ROUTINE

INTRODUCTION/PURPOSE

The app-based commercial vehicle carrying washing machines for washing purpose includes a drier and pressing facility will be available regularly or on order via the app.

- The idea behind this is to give the household women a free and fair life; here comes the concept of gender equality and men who believe in this will become a part by contributing some of the weekend time by working and sharing the women's work by washing, drying and ironing the whole week's clothes to be worn for all the family members and paying their respects to household women.

- The vehicle will have commercial or industrial-type heavy duty washing machines that are capable of washing and drying blankets, quilts, carpets and anything for that purpose, which are not possible to wash at home by women of the household, and even smaller size washing machines for washing clothes regularly.

- The washing machines will be solar or diesel-powered, along with the company's special washing powder (a special formula washing powder that is very effective/tried/approved/endorsed), which will be approved by the government or concerned authorities, and the powder will be sold to the customers at a price affordable and less than the existing so-called washing powders, which are not so effective in giving fragrance and shine and cleanliness to the clothes, blankets, carpets and quilts and killing germs, bacteria and viruses.

- A special team has prepared the washing powder, which is licensed, tried and endorsed by many for the past 20 years.

- The vehicle, which can be called a roaming washing machine, will be visiting the locality upon request with the help of an app and all the payments will be made after using the services.

- One more purpose is to create responsibility and teaching our kids to share the workload of the household.

- The main purpose is to give respect to the household women and not to impose and bind them to the household works, as it is even the responsibility of the men.

- Even the water conservation and power saving will be an additional feature as not more water is wasted comparatively.

- Instead of washing daily at home, it can be done twice a week and power and water can be saved and we can use the treated water or water good enough to wash.

DESCRIPTION, FEATURES AND BENEFITS

The purpose of these innovative ideas and improvisation is to create jobs for all kinds of people and preferring the Indian products and putting India forward when compared to products from foreign countries and this applies to all the chapters and ideas in this book.

Just imagine how many jobs will be created by saving water and avoiding health problems for women and each and every individual by making life easy at no extra cost but a lot cheaper comparatively than that of earlier times.

Imagine the number of engineers, technicians, app developers, drivers and cleaners that will be induced and will get more job opportunities.

- Instead of using the washing machine daily and wasting power, water and human energy, just switching to the concept and utilising the facility will make life easy.

- For people who move more frequently and can't keep on carrying the items in home shifting, it will help them.

- It will help even those who can't afford the latest washer and drier and can't pay excess power bills every month.

- The washing machine vehicle will have different sizes of machines as per requirement, which will be driven by a driver and includes a cleaner who will help assist and will do the ironing as well for extra pay and we can indulge in our other household works while the washed and ironed clothes will be delivered at our doorstep even if we are not present, or they can be delivered to the watchman of the particular society with a tag and description.

- The vehicle will be modified to fit in all the necessary equipment and even old models can be used and it will be modified to take the solar power and diesel power as well as battery power whichever is cheaper to run the machines and with less water.

- By requesting the time, it will arrive or else can wait for the regular roaming and visiting hours daily; these vehicles will be good enough to cover all the areas of any city and this applies to all cities throughout the world.

- Payment directly from the app and washing powder sachets will be given for self-washing or taking extra service and the delivery at the doorstep or to any caretaker if personally not present at home with pressing and ironing as well.

- All the technicians and engineers will see to it to modify the vehicle to carry enough water including the drainage system and enough machines and accommodate people for washing their clothes and the power generation by solar battery or diesel will be the main innovative part.

- Even the heaviest blankets, carpets and quilts can be easily washed with less effort, less water and less power by saving these when washed at home comparatively.

- These commercial washing machines at a maximum can take a load of up to 300 kg, which is impossible at home, and still wash the things at a cheaper cost.

- The male household persons can enjoy the work of washing clothes and saving power bills and saving water.

- Look at the jobs created, which are immense and in large numbers, for engineers, technicians, cleaners, ironing people and drivers and when the washing powder (SECRET BEST POWDER) is a hit among the users, even a production plant can be opened or can tie up with any manufacturing plant on a profit-sharing basis.

- Finally, I would like to say that improvisation and innovation are the backbone for a stress-free life and comfortable way of living life; we have been neglecting the hard work done by our household women for years and decades, it's time to share their work and keep them healthy and happy.

CHAPTER 4

SALON ON WHEELS, VANITY VAN CONCEPT

AN APP AND A COMMERCIAL VEHICLE AS THAT OF A VANITY VAN FOR PERSONAL GROOMING AND PERSONAL FASHION DESIGNER ON DEMAND AND EXPRESS SERVICE AT A CHEAPER COST

INTRODUCTION/PURPOSE

- The main purpose is to bring all the comforts enjoyed by the privileged, rich, celebrities and famous personalities at a cheaper cost available at the doorstep to all the classes of people.

- Why only the rich and privileged should have all the comforts when they can be available at affordable prices to the rest of all classes and masses?

- So, basically, a commercial vehicle will be modified into a vanity van, each separate vehicles for men and women, one of its best kind of features, which can accommodate 10/15 people at a time with all the facilities as that of a 5-star saloon, a washroom and shower with a personal fashion designer tailor with express service and the dresses will be stitched by the time we finish the activity of cutting and shaving or for women make up and massages and activities and will help in getting free advice while going to a party and can select from all the available varieties of dresses/handbags/shoes/sandals/chappals.

DETAILED DESCRIPTION

- Well-trained beauticians, stylists and tailors will be present in the vanity van who can at a time serve 10/15 people with all the features and facilities like a washroom shower, washbasin, pedicure, manicure, massage, full body massage, oil massage, sauna steam barber, etc.

- We can order for service and instantly get the charges through an app and if suitable and affordable, it can be available at the doorstep and the screens inside will be showing the latest fashion trends and fashion shows and the same clothes will be available with the fashion experts including a tailor which can be stitched by the time we finish the activity.

- It will be separate for men/women/transgender.

- The main purpose is to give all the comforts and services available to the common man at a reasonable price and give them a king-size or queen-size life.

- The payment will be a yearly package to avail all the services at a cheaper cost 'N' number of times, but pay as you use will be expensive and the service will be open for running customers as well.

- Wherever there is a marriage or reception or any happy occasion, the services can be requested and the vanity van will arrive at that particular location, and the whole day, the services can be utilised after every ritual, changing the clothes and hairstyle and makeup and bathing shower and all salon-related services will be available and if it is not enough, extra vanity vans will be requested to serve each and everybody as per the requirement.

- A yearly package or yearly card can be used for the whole family.

- A cardholder can use all the services and for every individual, the payment will be individual or yearly package and one can avail the regular services 'N' number of times.

- The jobs created will be enormous like the jobs in vehicle modification and giving an opportunity to girls and women by giving them beautician and tailoring course and even the latest fashion trends will be followed and sold like hotcakes and even the brands/fashion stores can display their fashion statement.

- Bride and bridegroom packages available at reasonable prices and will start the course well in advance of the special occasion and continue till the last day of all the functions and parties.

- People attending interviews or auditions or people travelling from faraway places to attend functions and marriages or going out of the station just to attend the event can request for the service and get freshen up before the occasion/meeting/schedule.

- People coming for auditions instead of buying expensive clothes can rent expensive clothes or wear for some occasions and can return, what's the point in buying expensive clothes when not worn regularly.

- The main purpose of the 'salon on wheels' concept is to bring the celebrities lifestyle to the common people doorstep at reasonable prices and yearly packages and all the household will enjoy the services and benefits.

- The designing and the modification of the vanity van will keep on improvising by the demand and new set of vanity vans will be launched yearly.

- A lot of companies have launched this concept, but once launched again with greater force and with newer features, benefits and services, this will definitely be a hit.

- Even the same can be used for giving the training to newer candidates in the fields of grooming, salon, beautician, tailoring, fashion designers, etc.

Chapter 5

HEALTH CONSCIOUSNESS AND AWARENESS FOR MORNING WALKERS INCLUDING SENIOR CITIZENS

INTRODUCTION/PURPOSE

The main purpose is to form a group of trained/certified dieticians/personal trainers by visiting each and every park guiding the morning walkers, by suggesting the exercises they need, how to reduce fat/weight and be healthy, also checking their bp and maintaining their health record, checking their weights regularly, giving morning juices when needed and including special services for people under any medical treatment, which all will be covered underpayment for the yearly scheme.

DESCRIPTION/FEATURES AND BENEFITS

- We see everyday morning walkers in parks and footpaths who are not aware as to how to control their health and what are the things needed to cope up with their medical conditions; this concept is created to help them, what they normally do is watch it on some health channels and do the same but what is required is proper guidance and how to get it right to maintain their health.

- The trainer is appointed to make them conscious about the diet they need and the exercises they have to do regularly, to avoid damages to their backbone pain and knees pain conditions and to guide them in buying proper shoes, clothing used and maintaining their health records including bp and pulse and physical fitness.

- It will be taken care of for an affordable price and will come by a half-yearly or yearly package scheme; even monthly trial will be given for a month's price.

- The purpose is to guide each and every person to do a proper workout and despite their medical conditions and weaknesses give them a healthy and stress-free life.

- The special team will visit each and every park in a vehicle every morning to start the session along with necessary gym equipment, vegetables, fruits, for preparing juices and serving, which will be included in the yearly package.

- In some emergency cases, give them CPR, wherein till the ambulance has arrived, because they are not available when in need.

- Make a record of daily bp readings, weight control tips and exercises, diet tips for obese people, and smart workout procedures rather than hard exercises.

- Celebrity visits for creating awareness monthly once or twice and free health tips from health consultants and suggestion of visiting doctors are good for their special conditions.

- A team of 5/6 people will handle a crowd of 40/50 walkers by suggesting diets, exercises, special straps for body and bone muscles, and suggesting shoes and clothes etc., and giving them daily juices as per their medical conditions, providing mats for exercises to be done while laying on ground and sitting, and giving motivational speeches and advises.

- In today's busy world, the senior citizens are the people who are mostly neglected and paralysed patients are not taken care of and how they are left on their own without support even if they have the will to overcome the shortcomings and win over their condition, the team will give hope and respect and listen to their problems and give them the motivation to overcome the condition by taking care of their daily routine and exercises and give them a purpose to live a happy life.

- The team will carry yoga mats, chairs, gym equipment and big balls filled with air and show them how to do aerobics for back pains and joint pains, which is a total health fitness package with simple stress-free exercises.

- The team will be deployed only in the morning time and the volunteers and employees will get their salaries for this part-time job and can continue with their regular jobs in the day time.

- Even training will be given to the interested people, for girls and women who want to become volunteers and who are from the same background or any other background.

- This concept will give extra earnings for those interested and in return, the customers can get a better and healthy life.

- The gym service in the locality will be open for one day in a week for all these people which will be called the senior citizen special day when the gym is not busy on the weekdays as a courtesy and respect towards the elderly in our society, thus giving publicity for that particular gym in the area which will attract more customers by doing social work.

- The volunteers will have all the information of the doctors available throughout the city for their medical conditions at a less and affordable price or special discounts for the senior citizens.

- Monthly celebrity visits will be on the day of marathon run and conducting sports and giving out prizes for the winners and motivational gifts to encourage others.

- Finally, senior citizens, elderly dependent members of our family, are like the backbone and always waiting and wanting to help us out even if they are not capable of, they give us hope and blessings, shower love upon us and our children and in return expect some love and hope for living, love for their heart and soul, and grandchildren are like their friends, so let us create some atmosphere of love and respect by sitting near them and talking to them about their daily routine and the improvisations they want and how they want to be treated and just by listening, all their problems will disappear, they just want someone to listen to and make them feel they are important and the backbone of our family and society.

- Father is the door of heaven and mothers feet hold heaven.

- At the least, we can support them by motivating them to maintain their health instead of the whole day asking them to just do household works and laying on the bed.

Chapter 6

ROAMING CONCEPT (VOLUNTEERS, APPS, CAMERA), WORLD TOUR, VIRTUAL TOUR FOR ALL FINANCIAL BACKGROUNDS, ECONOMICALLY WEAK/STRONG INDIVIDUALS/GROUPS

INTRODUCTION/PURPOSE

Developing a camera/video recording with all four sides along with an app for requesting a visit who are ready to volunteer and by bidding the price and quoting the lowest with utmost service will be volunteering the visit.

The purpose behind this concept is

- If any individual need to visit a place, religious holy site, friends, relatives in any part of the world and want to avoid all the stress of travelling or cannot travel because of any reason, they can just hire a volunteer by putting up a request in the app and the persons/volunteers from that destination place will bid their price per hour and whoever/

whichever is the lowest and gives best service will be accepted for that visit and he will be volunteering the visit and get the work done.

- Even shopping can be done or purchasing a gift and delivering to the loved ones or purchase an offering to the deity and delivering.

- Or visiting any place to refresh childhood memories by visiting their places of request.

- Can request a visit to any place and the locals from that destination place will take up the request and fulfil their demands for an hourly price.

- The hourly price will be standard throughout the world or less than that as per the volunteer's agreement.

- World tour concept without travelling in actual and still be able to view with four cameras recording all the sides and giving live feed to the requester in the 4D screen in ISLEEP as well.

- Visiting places, which are physically/financially not possible, in the lifetime.

- A volunteer will be hired on an hourly basis or for a full day and will be transmitting the live broadcast with the help of a special camera, four sides recording and transmitting accordingly on video call.

- The app will help and allow to pay the volunteer only after the service is given and visiting all the places' requested payment gateway.

FEATURES/BENEFITS/DESCRIPTION

- An app will be developed where you can register to be a requester and/or volunteer simultaneously.

- A camera video recorder will be developed, which will be purchased by everyone, 4D concept recording, broadcasting and transmitting.

- Hiring a volunteer for visiting places of request, religious places, university work abroad, visiting and site seeing around the world, shopping for items which are famous in that particular part of the world and speed delivery by booking air shipment/courier.

- This app and concept are linked to the smartphone or ISLEEP, which will be the platform to watch a video live or recorded with all the special effects added, simulation effects, all the weather conditions can be felt including the movements of the vehicle speed breakers even the slightest movement can be felt like that of the original recording.

- Can request a visit to any part of the world and will come up in the app and the willing volunteers will respond and finalise the time slot and amount.

- Can visit seven wonders of the world just by putting up a request.

- Can even request to visit far off relatives, friends and even visit ceremonies, functions and even present them the gifts purchased with the help of the volunteer.

- For many, the world tour is far from reach and a never fulfiled desire, but with this concept, any individual from any financial background can put up a request for a visit to any part of the world and fulfil his lifetime wish at a cheaper cost comparatively, for e.g., if someone from turkey wants to see and visit India virtually, then a formal request is raised and the volunteers from that area will respond and give their best prices hourly/half day/whole day and even the requester will put up a mode of transportation, public-private cycle, bike, car, etc. and will mention the services needed like shopping, packing and shipping any liked or purchased items and it will be done through branch offices worldwide in a fastest and quickest way to reach the destination and it will be the safest way, so by hiring a volunteer, he or she will travel to the desired places, even assist in shopping and bargaining and helping in packing and delivering the items and it will be delivered to the doorstep of the requester. Finally, the charges will be credited into the account of the volunteer once the service is done and in the app itself, it will be visible instantly, through a payment gateway, which will be secure for both.

- This concept even helps international students travelling abroad for studies and want to be familiar with the places and bus routes and even search for accommodation with the help of volunteers even before visiting the place.

- This concept helps in knowing the world and becoming familiar with the latest trends around the globe.

- Elderly senior citizens can request a religious visit to their favourite destination, which can be a live broadcast or recorded with simulation effects and can be viewed in ISLEEP.

- The world tour recorded videos will be available from an expert, who will visit all the places from any place to any place by any means and is readily available for rent or purchase with all the simulations and 4D effects of climate, temperature, ambience, odour, etc.

- Every volunteer will put up a video of all his/her visits and will be available for rent or purchase, whatever it is like a boat trip in Italy or bullet train journey in japan or boarding a double-decker bus in London or horse ride in India, it will be in detail how to travel from one place to another covering a whole city and all the modes of transportation will be covered and even will explain the best and fastest way to reach.

- The volunteers will record all the videos and explain in detail about their place and put up in the app, the important places, the busiest markets, the shopping markets and what is their place famous for and what can be purchased at a cheaper price as compared to any other part of the world; each and every street will be covered and all the visits to different places will be recorded and will be available for renting or purchasing and with time, the videos will be updated.

- So with this concept, every individual in this world is familiar with the rest of the world at a cheaper cost and will get to know the rest of world, their culture, the present situation, food habits and even the food can be packed and delivered to any part of the world in hours with the help of express delivery in case any individual likes any food of any region.

- Even the hot food prepared can be cargo delivered to any part of the world with express service so the concept of a global village comes into the picture, thus reducing the distance between nations and humans without borders and nations, without any restrictions for visit.

- Visit the sick in any part of the world, visit the far off relative wishing them on their happy occasions, taking part in their celebrations, functions and even the food can be packed and delivered for us to be a part of the celebration, so the visit is virtual but true participation is possible with this concept.

- Even a surprise visit or an appointment visit is possible.

- The main concept and benefit behind this are even if there is no one known to any individual in a country, one can still be able to visit any country and do a virtual but real tour and enjoy his/her time with any means of transportation from any place to any destination except restricted places like intelligence or military places, even can visit famous liked celebrity homes.

- Finally, the concept works both for domestic and international travels, even in the same city, by hiring a volunteer and he can go all around the shopping places and whichever is the best place or favourite items found at a particular place, the customer can himself go directly to that place and purchase instead of running here and there and finally getting tired and postponing the shopping.

- The advantage of this app is that we have the right to see what we want to see and visit wherever we want even if we don't know anybody there.

- In the rest of the apps, only watching is available but this concept gives the power in their hand and they can just request for the visit and instantly it is done and that too the live visit.

- Like if I want to see white house then the request is put and the volunteers in that area will communicate and show the live videos and we can enjoy the sight.

- Sometimes, the old and bedridden want to see their childhood places where there is nobody in contact, but if their desire which would be their last is not fulfiled and is because all are busy or don't want to go to that place, but with this concept, just put up a request and a volunteer will be at service.

- Such a wonderful concept by utilising the video call apps and giving them the live feed.

CHAPTER 7
W/C TRANSFORMATION FOR BETTER TIME QUALITY AND ENJOYABLE

MOVABLE W/C PLATFORM, SCREENS ALL ROUND, FOR RACING CAR AND WATCHING SCREENS FOR VIDEOS PLAYING FOR VARIOUS PURPOSES

Introduction/purpose

For centuries, the w/c concept or to attend the nature's call has been evolving, and in the past few decades, the usage of w/c for attending the nature's call has been boring for all age groups, so I want to introduce the new concept of w/c wherein it is on a platform in the washroom and is surrounded by screens which help for better relieving and it is movable all around the washroom with self-wash jet spray/fragrance shampoo/perfume spray for better hygiene.

The purpose is to modify the w/c and to make relieving more enjoyable and satisfactory and farting and screaming easier and inaudible as in a lot of people are habituated while relieving, without even disturbing occupants of the house or someone nearby to the washroom.

The washroom walls will be panelled with screens which will be waterproof and break-resistant and will be the screens for watching videos and playing games; by sitting on the toilet seat for relieving one can play a racing game, watch the news while rotating and revolving and moving within the space turning and going back and forth by holding the handle as that of a car steering or bike handle and even pedals for leg rest and will work as brake and accelerator and the whole concept will be changing through this modification.

BENEFITS/FEATURES/DESCRIPTION

- One can relieve easily and stop worrying about farting and screaming and making vulgar sounds, which will leave one embarrassed.

- Senior citizens face difficulties in relieving themselves, and medically for better health, it is an important thing to clean out one's stomach every morning without constipation and with ease.

- This will be liked by all age groups especially kids/infants who are to be taught to sit on western w/c and make it enjoyable.

- One can sit and relieve and at the same time, move around like the electric car which we see in the exhibition, malls and play zone and can relieve in very less time.

- Instead of sitting idle while relieving, one can watch news or documentaries or play games or students going for exams can revise and the screens will be connected to the data and all networking sites without a camera and smart TV connectivity, apps, etc.

- The drain system will be vacuum suction type and self-wash jet spray, shampoo spray and freshener spray will be fitted as per the requirement.

- The seat will be height adjustable and size adjustable as per the butt size and comfort sitting.

- The handles and pedals will have all the controls like that of a car and bike as per the game requirement.

- Any other logic games or for elders can be loaded from the manufacturer or new games designed by me will be available.

- The lighting ambience and ventilation and the amount of exhaust and fresh air supply will be done by air balancing.

- The seat height and seat size width will be playing major roles in this modification and with pest control sprays, vacuum and suction technology will be used and one urinal line from ISLEEP will also connect in the separate drain with same suction and vacuum technology, which is explained in ISLEEP.

- The w/c seat will have extra features like a steam facility for face and head massage and oil massage before a shower.

- Finally, the more we are freely relieved, the more we are healthy and to avoid constipation, we can have different sitting angles and the w/c seat can even perform manoeuvers, which allows relieving more comfortably.

- Games for all will be from the manufacturer and even the settings will be set/memorised by the system of AI and will adjust when an individual will enter the settings choice of 1, 2, 3, etc.

- The concept is to improvise the things around us and make it more enjoyable, hygienic even if it is a w/c.

- Have you ever imagined a w/c becoming the car simulator and the software will have all the models of cars for racing and enjoying and even flying or boating whatever depending upon the choice, will be loaded?

- A general universal problem is that kids and children without some assistance will not and can't sit in w/c for long and the results are known, so to make their using the w/c more enjoyable, relieving and self-cleaning all are customised and give good health results.

Chapter 8

100 CITIES
100 PROFESSIONALS

HOTEL/OFFICE AT THE SAME LOCATION WITH CHEAP LODGING AND AVAILABLE AT EVERY HIGH EXITING THE CITY WITH TRANSPORT FACILITY TO NEAREST METRO AND USING THE APP FOR CHECKING IN

INTRODUCTION/PURPOSE

- 100 professionals 100 cities (engineers, doctors, chartered accountants, technicians and defence personnel retired) whoever wants to come together and work, but on one condition, should have a vision and passionate about innovations and based on their experience should be able to give a real solution to problems around in the society or should be able to help in creating apps.

- A chat group can be created and can have meetings and discussions for positive results if not able to meet daily and can decide on the investment and execution.

- 100 professionals from 100 cities will make 10000 professionals.

- These 100 people will form a company and will invest 1 lakh each, which comes to around 1crore, and will work/chat/discuss in the evenings at a commonplace wherein the hotel or lodge will be opened for tourists at a cheaper cost.

- They will have regular jobs in day time but to follow the passion and to make the vision a reality and have the urge to create something useful and have innovative ideas, they will work/chat/discuss in the evenings and come up with hundreds of new ideas every day and do the R&D daily.

- They can rent out a place with some of the investment (part investment) at most regular busy outskirts highway and at the same place, they will come together to work, the rented place will be let out for boarding and lodging for tourists at a cheaper price with an attached central kitchen and self-cooking facility as well as a personal cook at a cost.

- An app will be developed for bookings for the customers and will have free cab service to pick up and drop to and from the nearest metro station.

- The rented place will be a hotel and a part of that will be used for office purpose sufficient enough for all the 100 professionals to come together and sit across and discuss and work.

- The hotel will generate funds for return on investment and to cover the overheads for office and staff salaries and utility bills payment.

- The hotel will be for families, individuals, bachelors, professionals and business personnel with separate entrance and exit ways to avoid unnecessary stalking and unwanted issues, with kitchen facility for all, either self-cooking or with cook facility along with the cheapest lodging in the city.

- If families are on a vacation or pleasure trip or religious tour or to visit any function of a far off relative and are on a long tour and want to avoid outside food, they can themselves cook for the whole day and a part of the kitchen will be rented out in the central kitchen with all the items and gas and burners and sink facility etc. and still cook in the early morning and take a parcel for the whole day if going on a sightseeing tour.

- This hotel boarding/lodging business will help all the investors in running the office premises because they will be working for free initially till they are well off to cover the overheads.

- Each city throughout the country will follow the same concept.

- Visionary, passionate, innovative individuals who want to make a mark and leave an impression will build an empire out of their contributions, which will last long for decades and centuries and will carry on a legacy forever.

- 100 cities, 100 professionals and 1 lakh investment each will invest 1 billion in total in one shot and imagine how drop by drop will create an ocean, but the main purpose is to feel the power of coming together and working together for the same cause and betterment of the country and rejecting the trend of getting obsessed with foreign technology, wherein they have the talent to reach to unimaginable heights of success through their vision, passion, innovations and ideas.

DESCRIPTION/PURPOSE/FEATURES/ BENEFITS

The people across the country who have a passion for innovations, R&D, improvisation in different fields, creating useful applications and apps, and are qualified and experienced from different backgrounds, who have a vision, will come together for a good cause and contribute their time, energy, money and self-indulge in group discussions through group chat and do research and development about different possibilities and apps creation and innovations and by copyrighting their thought will make it a reality and the hotel will take care of their part-time job expenses and profit sharing will be done if business is good as this is not their primary business.

The property can be leased for a very long time or purchased because this will be their permanent place to run their setup.

The main guidelines will be

- Taking up even the smallest job at a minimum and competitive price for engineering jobs designing or drafting or execution estimation in any streams like civil mechanical computers software developments, etc.
- The main purpose is to make a name of their own like the wonder team or dream team, the professionals and people will have trust in them

in the quality of work or trading or delivery or any service they want through their contacts even getting the building materials at a cheaper price or plumbing material at a cheaper price directly from the production plant or else execution of any commercial building or residential building or industrial building or any service of electronics at home or making a team of all trades like carpenters, electricians, plumbers and forwarding it to the people who are in need at a cheaper price and having knowledge of all the problems faced by the people and giving a solution to them at an affordable price.

- R&D of own country products and endorsing it on demand by the manufacturer, which normally is given to some celebrities who are not even using the item just for the sake of huge sums of money they endorse a product and have minimum knowledge about the product and never actually use it.

- Regular R&D will be done in all products and all fields and all topics and will have full knowledge of each and everything, benefits and uses and what are the harmful things widely used without the knowledge and creating a website and putting up all these concepts and what they do and what are the services given and what is their vision and what they are passionate about and giving instant solutions to almost all the problems and how they are planning to flourish and make a 100 billion-dollar company in the next few years and people will get to know the true value of engineers and will encourage them to do their best.

- They can take civil, mechanical, CA, electrical, plumbing, interior designing, whatever job and give the services at the lowest price and give additional services like material delivery at a cheaper cost with the influence and contacts of their group.

- The main purpose of the group will be innovations and bringing out ideas and will get a platform for doing R&D where any individual cannot do on his own if alone.

- The purpose is to make the country stronger financially, economically, technologically and endorse home country products, not to get obsessed with the foreign technology and products and items and reject the thought of looking on to some foreign company as ideal.

- People willing to support home country will definitely be obsessed with the thought and will give the work.

- Each and every job will be taken up and people with special skills can be hired and assisted in finishing the job in every field.

- They can endorse the products, create, produce or fabricate and can manufacture even from a small soap to washing powder till rocket technology and can tie-up with local companies who are already in the field.

- People who are willing to rule the world will contribute their time and money and in a short period will become the BHARAT-MEN in the field of technology and apps creation and innovation and rule the market in every field.

- No one individual will be a boss or an employee; everybody will hold the same position as a visionary.

- Once the app or innovation is a hit, then profits will be divided equally and they will never ever sell their company as this will be their identity as the masters of the world in every field.

- Proper confidentiality deed and partnership deed and any agreements for that purpose will be properly created and documented and the group will be famous and called innovative intellectuals.

- Business can be brought at initial stages from their contacts and targets can be achieved and can easily spread in the market and can be known for their commitment and service and passion.

- The main purpose of this group is to make a multibillion-dollar company within a short period by innovation, invention, services in different fields like medical, building industry, accounts, household products, hospitals, hr, training, development, etc.

- If more number of professionals come up to join, then they can take up the new location with another set of 100 professionals at another highway and hotel and the same concept but throughout the country, the 100 companies will be working in conjunction.

- 100 companies, 100 cities can go up to 200 cities and more depending upon the no. of people interested in innovation and following their passion.

- The same hotel concept can be followed and the design and how the hotel will run are in detail with me, and how the app works for the tourists visiting the city and how they will be guided and how to inform well in advance and what are the benefits of staying in this hotel are in detail with me.

- The hotel concept is to give the tourists peace of mind whenever visiting a new place where the other hotels rip them off the money and don't give proper services and the visitors are in constant fear of their belongings and food and sanitation and hygiene, safety and security of their family when they are out for some work.

- Finally, the concept is united we stand and divided we fall, one person who is a visionary cannot go through the process of getting funding and running a startup, so this group will help in achieving their dreams and a small amount of contribution from each will make wonders and likewise have the power to contribute in multiple ideas at the same time and try whichever succeeds will bring them lots of returns.

CHAPTER 9

APPS FOR ANTI/ E-COMMERCE GIANTS

INTRODUCTION

The app is to protect the small and daily street vendors who solely depend on their day-to-day sale for their livelihood.

The main purpose is to protect the business people around us and small and big shops and street vendors, poor middle-class vendors, needy senior citizen vendors and each and every street vendor from the dangerous foreign e-commerce giants.

Every area is surrounded by street vendors who are in real help for our day-to-day activities regarding our food daily essentials and commodities and for different items and their services, shops like medical, salons, phone recharge, vegetables, pan shop, tea and tiffin shop, stationery office supplies, puja items, florists, etc.

The app will be developed to protect our vendors and keep them alive in the market and business and support them financially.

The app will give information on all the vendors around us and their daily offers and discounts including their location, age, name, number of the vendor and how many dependents are being supported by this business.

The business in our country on the streets is run by courtesy, humanity and helpful nature. Otherwise, these vendors wouldn't have survived.

Nobody should support the foreign e-commerce giants eating away all the business of the street vendors and destroying them and their dependents if they overtake then lots of vendors and business people will be losing livelihood and even their families will suffer.

What's the point in making the rich richer by supporting them, wherein poor and needy are becoming poorer and poorer day by day.

DESCRIPTION/EXPLANATION

- The app will show each and every shop street vendor around us and will cover each and every area around our city, state and country.

- Every day, the daily discounts will be put up by the vendor and all the details of the vendors will be put up and how many dependents are dependent upon this business will be scrutinised and will be put up in the app and how much is the break-even for him/her to survive the day will be put up and at the end of the day, if anything is being wasted, it will be put up for more discounts for sale purpose and for fulfiling the target.

- At the end of the day, more and more discounts will be highlighted for consumable and perishable items along with sharing the live location of the moving vendor if they need some help for selling their items to feed their families.

- The main purpose of this app is to support our fellow beings, reject foreign e-commerce giants and support those who are striving for their livelihood to support themselves and their families.

- The vendors can register their business along with the location where daily they run, sit, stand and sell, and type of product and highlight their prices and discounts and delivery options and quality of their products.

- The purpose is to bring back the confidence in the vendors around us who trusted us and our neighbourhood and set up their business.

- Finally, in the end, they can highlight their sales figures and how much they are short of to overcome the losses, and seek help and assistance for improving sales so that they can sleep full stomach and feed their hungry dependents waiting at home.

- The purpose of this app is to support the poor working class who strive to keep them alive; whatever is available online is available around us and in the surrounding areas; so why to purchase the same thing? Instead, we can buy from our surrounding area and support those needy and hardworking vendors and daily street shops who set up their place and set off in the night, even when they don't have a proper commercial rented space for doing their business.

- This app supports farmers, street vendors and veg markets where poor people come to sell their items and feed themselves, so think how important it is to support these people and at least give them a source of living, who work and strive not to fulfil their desires but to eat and survive at least.

Chapter 10

ROBOTS CREATION, ROAMING PURPOSE

ROBOTS CREATION, ROAMING PURPOSE OF ROBOTS THAT CAN RUN, WALK AND RECORD VIDEOS AND INTERACT WITH HUMANS, MANUFACTURING AND PROGRAMMING, WORKS ON BATTERY AS WELL AS SOLAR POWER AND CAN BE CHARGED AT POINTS AVAILABLE AT LOCATIONS ACROSS THE CITY, MOST PROBABLE LOCATION UNDER STREET LIGHTS, AN OUTLET OF POWER SOURCE WILL BE GIVEN AND USING FOR ROAMING ON HIRE BASIS AROUND THE CITY.

USED FOR ENTERTAINMENT AND DELIVERIES AS WELL.

USED IN CRIME WATCH AND CRIME CONTROL AND SURVEILLANCE AND CAN ALERT THE NEAREST POLICE OR AMBULANCE SERVICE OR FIRE FIGHTING SERVICE OR ANY DISASTER MANAGEMENT TEAM.

CREATING A FEARFUL ATMOSPHERE FOR THE CRIMINALS AND ROBBERS WHO PLAN TO EXECUTE IN LATE NIGHTS.

WILL REACH PLACES WHERE POLICE CAN'T REACH EASILY IN THE DARKEST NIGHTS AND IN

RAIN, THUNDERS, DEEP FORESTS, STEEP CLIFFS AND MOUNTAINS.

INTRODUCTION/PURPOSE

Robots will be created and will be placed at all locations around the city and will be activated through an app and can be hired to move around in the city just for pleasure or delivery or roaming just like the roaming volunteers as per the requirement of the customer.

The robots will have cameras all around and the robots will be able to roll and move at speeds and will be strong enough to resist any impact or will have shatterproof break-resistant glass around the camera.

They will be strong enough to withstand impacts from attackers, most probably criminals who want to damage when caught in the act of nuisance and burglary or breaking into any private property or causing damage to ATMs, and all the illegal activities.

Any individual who won't get sleep in the night can hire these robots and roam around the city with a live feed from the recording of video and transmitted to the ISLEEP, which is explained earlier or can be viewed on a smartphone or smart TV with all the four views as discussed.

They can be used for food delivery at night who are hungry and know the food joints which operate in the night time permitted by police to serve the late-night workers (call centres and cab drivers).

Once the work is finished or hiring time is up, then they can be left at any nearest charging point and the robot will get engaged in charging and gets self-locked.

The live recording will be transmitted to the customer, can be watched all around 360⁰ and can allow interaction with the people in the place where a robot is present and recording.

Surveillance and feedback by alerting the police or any other emergency services will be an in-built software in the robot, some code words or SOS calls can be identified by the robot, screamings from girls/women in distress and need help; they can be protected by identifying the sounds, which will be fed in the robot.

The robots will be like mobile patrolling and surveillance and will have charging points across the city under the street light point.

DESCRIPTION

- The robots will be activated through an app and payment can be done through the app and can be hired on an hourly basis or whole night or whole day depending on the requirement of the customer.

- Can be viewed from the ISLEEP machine and can transmit live videos at all the four angles into the four screens of the ISLEEP.

- Robots work like anti-theft, anti-burglary and police alerting technology to bring down the crime to a minimum and create a peaceful atmosphere.

- It is like private surveillance and can be used/utilised by any individual and common people.

- Reduces burglaries, helping anyone in need and alerting the police by reaching to those place where the police can't reach easily.

- Roaming and watching nightlife on roads and to spend time and keep an eye on your business places, which are far away to safeguard your place, or people who can't sleep in the night or else fun for kids by visiting places around the city and enjoying the view.

- Charging points will be available at all locations around the city and in the future, they will be used for charging the electrical cars.

- Works on both power source of battery and solar power.

- Interaction with people by asking for id and taking snaps/saving in memory in the late night, if in doubt.

- Can hire permanently at night time to watch the business place like a jewellery store or any valuable business place like a car showroom, clothes store, etc. wherein the CCTV sometimes can be manipulated or else the power connection can be cut and the burglary can take place.

- Food delivery can be done at late nights.

- Creating a safe and peaceful atmosphere and creating fear in the minds of criminals.

- Helps in bringing down the crimes and atrocities against women and girls who work late at night and walk home.

- Any accidents, mishaps and emergency situation can be reported to the nearest helpline.

- Even can travel a long-range with intermediate charging or changing robots at every transition point.

- In the day time, it can be traffic police and CCTV as well for traffic violations.

Chapter 11

VIDEO CAMERA CUM MAIN DOOR

A REINFORCED DOOR WITH BUILT-IN CAMERA (memory for a lifetime) ABLE TO RECORD BOTH INSIDE AND OUTSIDE THE DOOR

INTRODUCTION/PURPOSE

The purpose of the video camera, which can record both inside and outside, is to have surveillance as well as record all the daily activities of our loved ones in the living room or front room where normally the family stays most of the time and sit together.

When our loved ones elderly people are gone forever, we miss them and will be thinking if we had recorded their daily activities, it would have been better, but we already did by installing the video camera cum main door.

We can watch how the kids have grown up when we watch the videos after decades.

The video can be online accessed and watched from any place.

This leaves memories even after decades.

This will have memory unlimited for storing the data.

PURPOSE/BENEFITS/EXPLANATION/ FEATURES

Just fix it and forget and it will be running on and on for years till the kids grow up and reach their puberty and finish studies and marriage day arrives and their kids arrive, and then when they watch their life, it will be the most amazing thing to look how life has passed and how they have grown up and how loved ones have left and how they used to spend their time and how they played with their grandparents and how life has passed like the sands through the hourglass. This will be the best video to watch a story of a lifetime and three generations growing together.

Can be watched on smart TV or smartphone whenever needed.

When anybody loses their loved ones, the pain can't be explained so this will heal all the wounds deep in their hearts, the refreshing and happy and sad moments and daily routine of them and their funny and serious faces and happy and sad moments and angry and frustrating moments, altogether a life, a story, a feeling, a moment and a joy forever.

These videos will be available for future generations even after centuries and will be available for them to watch and to know how their ancestors spent their days, their nights and how they looked and how they behaved in the moments.

The memory will be unlimited, which can record even for centuries without any spam or virus or fear of losing any data or fear of hacking, a foolproof system will be installed.

Even the live recording can be viewed on a smart TV for any occasions and any events and any functions and any special celebrations.

Records all the things and goes on and on, such a system will be installed which will have night vision and motion sensor and will concentrate on the part where there is motion or else will run normally.

Chapter 12

TRANSPORTATION WITHOUT TRAFFIC JAMS

ONLY SUVs, EXPENSIVE SUVs, ROYAL CARS AND HIGH LUXURY CARS TO ATTRACT THE CUSTOMERS AND WILL HAVE YEARLY PASS, CAN PICK UP AND DROP FROM ANY POINT TO ANY POINT

NO OTHER VEHICLES WILL BE ON THE ROAD AND EVERYONE WILL LIKE TO TRAVEL IN THESE PRIVATE YEARLY SCHEME SUVs.

AN APP WILL MANAGE ALL THE PICKUP POINT CHARGES, DROPPING, PAYMENT, ETC.

INTRODUCTION/PURPOSE

Running SUVs in the city from houses to offices and regular pickup services and dropping services but in only royal and high-end luxury vehicles to reduce traffic and make a habit to the people to make use of this service and fulfil their desire and finish their work.

The package will be yearly and will be cheaper than the public transport and instead of each person taking his/her vehicle and creating a traffic jam in the city, they can utilise the service and enjoy a royal ride.

They can book from any point to any point and the amount will be charged as per the distance travelled.

People who can't afford royal cars or high-end luxury cars will utilise these services and enjoy a royal life for going to work or any other purpose and can go from place to place.

This service allows individuals to cut down the cost of their own vehicles maintenance, fuel break down, flat tire and all sorts of problems.

Security while travelling with family or even single girls or women can travel even in day or night without fear.

This service can be utilised by regular office going people, college-going students, ladies going on shopping and business people travelling around the city more frequently in late nights.

DETAILS/BENEFITS/FEATURES/DESCRIPTION

Yearly package for regular travellers at a cheaper rate, very less compared to public transport.

SUVs throughout each and every city, from any point to any point only booking through the app and payment through the app and distance will be recorded in the vehicle and the id will be read in the car and the main purpose is to bring down the traffic jams and give way to emergency services and quit by travelling all the other means of transport and enjoy the royal ride, Security for all, no need to walk to the nearest bus stop or metro or auto point.

The main purpose is to release the pressure of the envy and jealousness that only the privileged can travel in luxury cars, even the profits are very less, then the advantage is the roads are empty and free of traffic jams, noise pollution reduced and safe travel with safety features like live recording/tracking and hooter/flashlight/strobe light options to alert the onlookers if drivers create a problem for (passengers).

Why can't a common man enjoy all the benefits of life as that of privileged and celebrities?

Chapter 13

RELIGIOUS PLACES LOCATION

RELIGIOUS PLACES LOCATION THROUGHOUT THE WORLD WITH 24 HOURS AUDIO RECORDING AND TUNING IN FACILITY AND AVAILABLE FOR ALL IN AN APP

INTRODUCTION/PURPOSE

An APP is developed that shows the religious places specifically and live broadcast of audio for listeners throughout the world for tuning in and accessing speeches and sermons and mythological stories which are going on at that place or any other place and even earlier recordings saved so that no speeches and ceremonies are missed even if busy with other ceremonies.

The app will help visitors arriving and can easily locate their choice of religious place and can get a list of programs and ceremonies and speeches for the next few weeks, any small big or very interiorly situated places can be located.

The app will give all the information about the deities, the special ceremonies going on and all will be able to hear the sermons and speeches and even the previously-stored speeches for the past few months and all the data are stored in the central server and can be accessed at any point by any individual any number of times helping them meditate, concentrate and connect to their inner self through the preachings.

Even the future generations can access the preachings, sermons and prayers at any religious place by any individual of any faith.

FEATURES/BENEFITS/DESCRIPTION/ EXPLANATION

Any individual travelling alone or with the family to a new place can get to know the location of their respective faith temple, church, mosque, Sikhs place of worship or any religious place, its speciality and details of the committee and number of believers visiting every day, etc. and even get the audio/tuning in accessibility for that place even from any place.

Simultaneously, all the sermons going on can't be attended so the saving option will help in listening to it at a later free time.

Any individual attending Friday prayer may miss his favourite speech at his regular place of worship so later in free time, he or she can access the speech and listen.

Even for the big nights celebrated in many faiths, the simultaneously going on speeches can be heard so the speeches are stored in a central server from where they can be accessed at any point in time.

The preachings are stored forever even helping the intelligence/national security personnel for reference.

This makes the preachers take extra care to not speak the forbidden and against the community and disturbing peace and harmony; only good preachings and good habits are taught and avoiding radicalism and creating an atmosphere of phobia of a particular faith or inciting violence, in any case, the preachings will be peaceful and good for the communal harmony and spreading love and friendship among all sections of the society.

The continuous recordings of all the things happening over there will be broadcasted/telecasted in live only audio and even when no speeches are going, one still can hear that place of worship and enjoy the silence.

The app will help in alerting the users for the prayer timings for that particular location and sunrise and sunset timings and fasting timings and facing their place of worship from any part of the world.

The app will give the information of any religious place, its images, images of the deities and the speciality of the place, the school of faith which it belongs to, the capacity of the place to hold visitors and presently how many are present at that place.

The in and out of the frequency of people like for Friday prayers, all the mosques are full so the app will help in finding the best timing to attend the prayer, no. of vacant places, the location and the timing of prayer and sermons.

The purpose of the app is to bring all the religious places under one app and to enjoy the freedom of religion and listen to whatever one feels like to any faith sermons without any hindrance from any group even without visiting that place and practice the faith of his/her choice and having access to the sermons and preachings of all the faiths at any given point of time and even available for future generations.

The app helps in practising good faith and getting only good from all religions and maintaining social harmony and friendly atmosphere in the society and even can have the access and information of the religious priest/head sermon giver or main lead of that place to clarify their doubts and interact without having to disclose the identity and their background and faith.

No doubt will be in the minds of the people that some extremism is being taught at other places of faith, communal harmony will prevail and anybody will and can follow any faith of his/her choice and can feel relaxed and will have no restrictions whatsoever.

If any objectionable or hate speeches are given, then the action will be taken by the authorities and can even have the right to ban that person or individual who is against the communal harmony and legal proceedings work will be eased as the proof is much stronger.

The app will give all the information of the religious place, who built it, how old it is and any aid is required or any marriages or ceremonies happening and what will be the prayer timings for the festivals and what are the facilities provided by the committee and how much funds is being collected monthly and how they are taking care of the expenses and the salaries of the staff of the caretakers and cleaners and head priest and deputy priest and water bills and power bills and if any orphans are being educated and any weaker sections of community are being helped by that place or any good social activity is going on and how they want to improve the relations with other places of faith and what are the future programs and all the information will be available in the app, any vacant portions if someone need cheap place to stay for few days instead of hotels and attend the preachings and finish their work in that city and finally can donate some amount for the development of that place of worship.

The app will create awareness and taking positives and taking good from all the religions and eliminating the doubt of someone giving the wrong preachings and hate speeches.

Any individual from any faith can have access to any preachings at any given point of time thus giving the freedom of religion set by the constitution.

The app helps even the kids, girls and household women to be alert about the prayers and timings and have access to the speeches of the place and become religious and observe fast and meditate and feel peaceful and do the household chores with interest.

The women who are not able to attend their place of faith can get connected to that place round the clock and even can have the access to speak to the main preacher and clear any doubts or ask for auspicious timings and days and maybe any other clarifications without having to depend on the other members of the family or without actually going to that place.

Any anti-social speeches will be alerted to the authorities and action will be taken immediately on any individual.

The app helps in going through the accounts or the financial status of that place and can donate to the place to support the salaries and utility bills and contributing to the expansion of the place even without visiting that place and can get all the information of all the donors and amount donated each and every day.

Any individual can have access to any place of worship in any part of the world at any given point of time and can enjoy the preachings and become a good citizen and good follower of the religion of his/her choice.

Any individual can follow the festivals of his/her choice, listen to ceremonies, funeral prayers, marriage ceremonies and Friday prayers and special night prayers and can gift any items to the priests and still contribute to the betterment of that place.

Any offerings can be given online and can get the offerings back via delivery at a charge.

Chapter 14

PUBLIC REPRESENTATIVE ACCOUNTABILITY AND BETTERMENT OF SOCIAL LIFE

GPL – GRIEVANCES TO PUBLIC REPRESENTATIVES FROM LOCALS

PRISON – PUBLIC REPRESENTATIVE INFORMATION STANDARDS OPERATIONS NEGOTIATIONS

APP FOR ACCOUNTABILITY OF MLA/MP OR WANT TO CONTEST IN FUTURE ELECTIONS AND CREATE A GOOD TRACK RECORD

INTRODUCTION/PURPOSE

An app which shows the details of local MLAs, MLCs and MP for public accessibility along with the contact numbers, for locals in distress or daily routine activities of cleanliness issues or power and water crisis or local nuisance from eve-teasers or any other issues or give a task for solving problems faced by the community to discuss issues and the app will be universal at any place and any country will show the required contact details pertaining to any issues and whom to contact and keep accountability of works done for the community by the public representative and have access to the accounts of funds given by the government and how it is spent and what are the future programs taken over by them and how they are planning to perform in the future.

Even any individuals can register their name and do the social work for the community and then get elected in future if they have a good track record and liked by the locals and the neighbourhood.

All representatives and future contesting representatives will get ratings as per the work done and problem-solving capacity and willingness to help the locals who have elected and how good and fast they have reacted and solved the problem unlike those who are out of reach all the time and once visible only in years at the time of elections.

They should be given daily tasks if found any issues pertaining to cleanliness, social issues and natural disaster if occurred and criminal issues, controlling burglaries, medical camps, social gatherings community works social harmony maintaining clean and green neighbourhood, helping local youths to get a livelihood, testing facilities set up in case of pandemics, necessary vaccines availability for infants free of cost from the government, roads repair and laying works, drainage and stormwater drain facilities and public toilet availabilities, taking care of the gardens and street lights and setting up CCTV in the neighbourhood and all social issues for a better and safe and happy life.+

FEATURES/BENEFITS/DESCRIPTION/EXPLANATION

The public representatives are accountable for the wellbeing of the local citizens of any local city/state/county/country in any part of the world.

The individuals who want to contest in the future for being elected can also get registered in the app and can maintain a track record and the ratings will be given by the people after every task and job given and how well they performed and everyday ratings and points will add up to the monthly and yearly performance and then finally at the time of elections and it shows the overall chances of reelecting next time or personal score of all the individuals.

The future contesting candidates can also register themselves and show their willingness and can show their performance how they cope up with the situation and problems given and doing social work, not for a photo-op but in real and helping the citizens and striving for the betterment of the youth and helping economically backwards and easily accessible to all and regularly updating the programs to be conducted for the betterment of the society.

When we make the public representatives accountable on a daily basis, then the betterment in the society and issues come up, which lead to the development of the community and in return, the country.

The app will have the details of funds allocated by the government and how it is being spent and the future plans and daily activities.

Any issues can be raised in the app and the responsible person will react and the particular department will take care of and close the issue at the earliest and finally, the rating will be given by the person who raised a genuine issue; all these ratings will add up to in a monthly and yearly manner and finally will give a record of activities done by the public representative and gets his personal score from 1 to 100.

Any issue can be raised and location sent and time slot will be given and can be seen how the action is taken and how swiftly it is solved and how much dedicated the public representative is actually.

If the problem is not solved, then automatically it goes to the next responsible representative from the escalation matrix given in the app, then even if the problem is not solved it will go to the next higher authority likewise it can go up to the ministry of state and at the national level as well and the ratings will be given accordingly.

Any person can register in the app for becoming the future public representative may be from any party or independent or ruling party and we can analyse the willingness of them to serve the community.

We have seen in pandemics and natural disasters how some public representatives have escaped and are out of reach when we are struggling to get basic necessities and they are safe in their holiday spots and farmhouses and are never accountable and accessible and enjoying on tax payer's money.

The purpose of the app is to create an environment of responsibility and create fear in the escapers from responsibility.

Any issues related to neatness, nuisance, loudspeaker noise pollution, potholes water stagnation and unsocial elements blocking the roads and functions blocking the main passways and roads and illegal stealing of power and water cleanliness, etc.

Each representative will respond accordingly and for each activity addressed, they will attach an image and ratings will be given from 1 to 10 wherein 1 is worst and 10 is excellent.

These ratings will add up to the track record of that individual for reelecting or losing.

Any individual travelling to any place can raise any issue and tag the person in charge of that area and can give the ratings as well.

Over a period of time, if the ratings are less for any public representative, then for sure, next time people are not reelecting him and any other person who is in continuous help of the community will surely be elected irrespective of the party or caste creed or background or maybe from the ruling party.

The public will give chance to a new face.

The app will give all the information about any area and any place travelling and will work universally and can access all the information and previous track records of the public representatives.

The purpose of the app is to create awareness about our rights and why we elect the public representatives and what are their tasks and how well we can utilise their services and have transparency in funds allocated for that particular area and right to information.

If each one of the public representatives is held accountable on a daily basis, then the whole country develops in no time.

Even the unemployed can utilise the service and get livelihood from them by demanding jobcentres across the areas.

All the areas can have job centres and at least all age groups and all background people can support themselves by doing part-time jobs or till they can get a good job.

A lot of services can be suggested by regular meeting and interaction between the public representative and community members.

Chapter 15

MOBILE TOILET WITH A/C AND VENTILATION IN A COMMERCIAL VEHICLE

SOLAR POWERED/DIESEL POWERED/BATTERY POWERED

APP TO GET THE LOCATION OF THE VEHICLE OR CALL FOR AT A REQUIRED PLACE

SEPARATE FOR MEN/WOMEN WITH ALL AMENITIES CLEAN AND TIDY WITH WATER SUPPLY DRAIN SYSTEM, CHANGE ROOM, SANITARY PADS AVAILABILITY AND FRESHEN UP FACILITY, HOT AND COLD WATER AS WELL

INTRODUCTION/PURPOSE

Designed a vehicle with solar-powered/diesel-powered/battery powered with w/c bathroom for shower bidet for women with music system radio and smart TV and Wi-Fi connectivity and all the necessary amenities.

A lot of busy people have no time for getting freshen up or go to a hotel and use the toilet and unnecessarily get embarrassed by asking for service to use the only washroom.

An app is developed which will give the exact location of the mobile w/c paid service and even can invite it to required place to peacefully relieve or freshen up for the next job or interview or audition or else attending a party after work.

While we use, we can request them to drop at the required place at an extra charge.

All the sewage can be dumped into the stp and the water will be raw water after getting treated from stp not consumable so that to save water.

DESCRIPTION/FEATURES/BENEFITS/ EXPLANATION

The main purpose of this innovation and app is to stop people from using unhygienic toilets and attracting diseases and defecating in open.

To help women and girls comfortably use the washroom without running to a restaurant and malls and buying pads and running here and there and getting frustrated.

The vehicle will be modified in such a way that the waiting area and the w/c both are the cleanest and odour free and will be having special fragrance and sanitation and anti-fungal anti-bacterial treatments after each use.

Even a vending machine will be available inside the vehicle and for the use of sanitary pads and disposing of those, and even the social service can sponsor the sanitary pads and soaps and all these services will be kept to a minimum charge to ease the citizens.

We have seen nowadays how unhygienic conditions exist in the public places of the urinals and toilets, wherein the disease carriers are mostly flies and mosquitoes.

It's a basic pay-and-use toilet but with a higher side.

Even relieving facility for the physically handicapped is the priority.

Overall, the purpose is to run a mobile w/c on the road, which will be like the king's washroom and queen's washroom and ease the strain of women and kids as in most of the times it happens.

The payment can be done digitally.

The maximum number of vehicles will be available throughout the city and the app will show the exact location with even the number of vacant w/c and how far is it from the place.

The most required and needed place is at the public transport stations and a lot of stressed travellers are fed up with the existing washroom conditions.

The visitors arriving at the city are badly in need of the washroom neat and clean and not able to find it, early morning visitors, late-night visitors and kids and women, they need a clean environment to freshen up with hygienic conditions.

Even the visitors who can't afford a place to stay can use and freshen up and get to the work and even return to their home place in the night without even hiring a lodge.

The main thing is to improvise the living conditions of the citizens and make their problems easier and a lot of people will come and sponsor this which in return will help them only.

Some wealthy sponsors can fund for each city and like this whole of the country can be covered and ease the burden and relieving ourselves can be fun.

Everyone needs clean and tidy washrooms and toilets but it is far from reach so by spending a bit extra, getting a comfortable place is not much.

Even someone visiting relatives instead of going directly to their home can freshen up and then visit them.

Using the washroom for bathing, showering and changing sanitary pads and freshening up all will be at a price less than that of hiring a lodge unnecessarily.

Even the good habits can be taught to our kids, how to relieve and freshen up, have breakfast and then visit the relatives.

This innovation will help travellers coming to the city for one day's work and then will go back to their home place; they can relieve and shower and go to work or interview and even without hiring a room in the lodge, they can be super fresh and go to their destination and finish their work.

This innovation will create more jobs for the daily workers, underprivileged people, who are begging on roads, they can live a respectable life and support themselves and give a better life for themselves and their children.

Even vehicle modification and maintenance will create more jobs.

The designing behind this will need lots of engineers and a production plant and all the infrastructure thus jobs are created easily and normal citizens spending money on useless pay and use toilets can use this service and the expenses and salaries are covered.

The vehicles will be available at every station and religious place and prayers place and a special vehicle for the performance of ablution can also be designed.

These vehicles will have smoking and non-smoking washrooms to ease the customer's requirement.

At an average, each city will need at least 300 to 500 vehicles, just imaging the earnings and jobs creation and at the end, the city is clean, and the roadside pay and use toilets can be rented out for some roadside vendors for their business places.

Even some function halls are full in busy times; they can hire for the evening and all the guests can be easily accommodated with the relieving facility.

Some restaurants have very limited washrooms, even these can be used there.

Hot and cold water as per the weather conditions will be available.

The most requirement of the mobile w/c is the government hospitals, private hospitals, function halls, colleges, universities, bus, rail, airports station and busy shopping areas and old city where it's always busy with shopping and not have enough toilets, roadside markets and bus pick up and drop point and throughout the city where it is easily accessible to each and every individual

Just imagine a vehicle will earn 2000 rs in one hour and at an average of 300 vehicles in the city and the w/c being busy for 18 hrs a day then the amount collected will be

2000 × 300 = 600000 per hour and 18 hrs will be 600000 × 18 = 1,08,00000

The one crore change is earned from the citizens in one city just by easing their problems and imagine how many numbers of people will be involved in the app development and vehicle modification and cleaning and driving jobs.

Imagine if the service is started in 100 cities, then easily 100 crores is generated daily and can cover the expenses of the salaries and overheads of the people involved, a great business out of the toilet usage with very less profit margin or no margin at all, just for the sake of creating jobs and giving a better and bright future.

Chapter 16

HOUSEHOLD EXPENSES MANAGEMENT TEAM

AN APP ALONG WITH THE TEAM TO VISIT UPON REQUEST AT A PRICE

PREPARING SCHEDULE FOR EXPENSES MANAGEMENT, HEALTH WISE PURCHASING OF GROCERIES AND STUDYING THE MEDICAL CONDITIONS OF THE HOUSEHOLD

INTRODUCTION/PURPOSE

An app developed to teach maintaining the household expenses and a ready team to visit upon request, and give the demo and explain the expenses control and money saving techniques.

The team will visit and teach all the inhabitants and occupants of the household how to purchase the groceries and have a healthy diet as per the medical conditions and explain in detail the purchasing and delaying of necessary and unnecessary groceries, electronics, clothes and any other items.

FEATURES/BENEFITS/EXPLANATION/ DETAILS

The app will be developed and will give the required information as to how to manage household expenses and save money, through a call centre, if required a paid service then the team will visit the place or home and study in detail and take the requirements of the household and take the eating habits and study the medical conditions and will suggest a plan and a diet good for all and how to control the expenses and maintain a healthy financial balance sheet.

Even will suggest the best cheapest place for the groceries purchase and through their contact, the groceries will be delivered at their place once in a week and look after their money management, even they will conduct a week-long class as to how to cut short the expenses and unnecessary expenditure and still lead a happy and healthy life.

The team will give any other valuable information and how to save water electricity and cut short the utility bill and what to stock and what to purchase on a daily basis and what not to stock and from where to buy the clothes and improvise their eating habits and food habits and expenditure.

Nowadays, everybody, even the best, needs advice as to how to save money still having healthy food and maintaining their expenses and cutting short the utility bills and avoiding outside food and saving money.

After a training and demo and a proper expenses control mechanism have been followed, the team will leave the place and when the family is really saving the money, then the appreciation and rating will be put up in the public domain through the app and can even share the details as to how much they have saved in one month, and if further help is needed, the team will visit and again and give a motivational demo.

This app helps all classes of society and even the rich and middle class and economically weaker sections and upper-middle-class and each and every individual in the society.

The team will be having a financial investment expert for short time plans and a person who is familiar with the purchasing of groceries and vegetable and meat and fish and chicken, a dietician and a person who knows the bank loans department and gets loans for solving their problems of transportation or purchasing a house or vehicle or planning a marriage within the household and how to purchase the needed items.

After studying if any special talent is found in the home, they can suggest how to earn extra income by teaching or utilising any other skill.

Even the medical condition patients will get help by getting the cheap medicines from a place where the medicines are cheap and given on the credit for 2 weeks or monthly, even purchasing generic medicines and its benefits can be explained and purchased from a trustworthy source.

All the aspects will be studied and the team will give a schedule and explain how the market and the ads and the companies cheat and make them buy and thus by avoiding the unnecessary shoppings and saving money will help the household in their emergency times.

The fees will be valid for six months and even after that they need more advice then can utilise the facility by paying extra fees.

If the family is planning any business, then the expert team will study the market and give them the advice good for all, and if the opening of the business is finalised, then they will assist in purchasing the interiors from a cheaper and best place.

The main purpose will be a foolproof planning and loss prevention and saving money at all costs, and if needed, making them interact with the people who have lost everything in the business and still coping up with the losses.

The team will readily give suggestions to any sections of the society and even for investors and people whose businesses are running in losses and how to cope up with their losses and can be mediators in selling the business at a service charge given by both parties.

Every now and then, the same business place is opened next to a place but both will be in losses, what's the point if there is no interaction or advice taken before opening that place and facing losses.

Every now and then, we have seen people losing all their and their parents' savings foolishly by investing in a place and can't run their business and want to escape by selling off that place at a very cheaper price, and then the suicidal tendencies grow and or else go in depression, and unless a helping hand is given to them and a push is needed to get them out of that situation and lead a peaceful life and need suggestion and expert advice frequently and monitor their financial and family life and become their well-wisher.

So the team will be a lifesaver, a money saver and even have the information of any recruitments or vacancies at any place and where trustworthy people are required and can help them settle down in life.

The team will be well familiar with that city and all the things available at places and know every bank official and have useful contacts and know all the lawyers and doctors and chartered accountants and knows about any vacancies and will help in recruiting the people and earning extra income by working part-time and extra work after a regular job.

The expert team will have all information about savings plans, life insurance, medical cards, purchasing four-wheelers, houses, two-wheelers, investing in businesses and getting share and percentage, and even where huge businesses are involved, help them with the market survey and what kind of business will be best in that area.

The team will have all the data of the special talented people and who want to earn extra and will help them in finding the right extra part-time job and some want some other help to come out of the situation even helping them with the loans and whatever required.

Even women who were specially skilled earlier and want to start again will get help in giving their skills to the next generation.

Finally, the team will have an idea of each and everything and will help in solving the financial problems of each and everybody at a charge through the app and even personal visit to give their best advice after studying the scenario of the situation of the household.

Chapter 17

WOMEN/GIRLS PROTECTION THROUGH AN APP

FAKE SWITCH OFF APP

THE MOBILE PHONE APP WILL SHOW SWITCH OFF AS NORMAL BUT WILL COLLECT THE DATE OF THE PERSON SWITCHING IT OFF FOR ANY PURPOSE OR REASON, WILL CAPTURE IMAGES IN SECRET MODE, RECORD VOICE IN SECRET MODE AND IMMEDIATELY SEND IT TO THE ICE CONTACTS AND EVEN MAIL IT TO THE CARETAKERS AND NEAR AND DEAR ONES.

LIVE TRACKING OF THE MOBILE PHONE LOCATION WILL BE SENT OUT.

EVEN A HOOTER WILL ACTIVATE WHEN THE MOBILE PHONE IS TOUCHED BY AN INTRUDER OR KIDNAPPER OR A BURGLAR OR INNOCENTLY ACTING FRIEND.

A SIREN OR HEAVY ALARM WILL ACTIVATE AND ALERT THE NEARBY PEOPLE, AND THE BATTERY POWER BACK UP WILL BE IN THE PHONE COVER ONLY AND WILL GIVE A BATTERY LIFE OF EXTRA 48 HOURS.

INTRODUCTION/PURPOSE

An app is developed especially for women/girls safety or any individuals who want to be safe and are privileged and rich and are unaware of the persons around them who act like their friends and maybe more dangerous at times and can do harm just for the sake of extorting money and by kidnapping.

A battery backup other than the regular battery will be in the phone cover activated only when the fake switch off is active, will give extra 48 hours for the phone life to be on and pass on the information to the ICE contacts.

This fake switch off app works when a stranger takes the phone and tries to steal by switching it off, the phone will act and bluff as if it is switched off but instantly starts to work in secret mode by sending the pics of the person and recording the voices and giving out hooter noise and the ICE will get all the information needed to get the culprit and alert the authorities.

The fake switch off will work in sync with the hooter pics and voice sending and alerting the people in the surroundings.

The switch off in the smartphone will be user friendly and will be modified as per the user requirement and with pin or passcode will be actually switched off, the smartphone will recognise the touch of the user and if any stranger touches it, it will give out instant hooter noise and alert the people around, the phone protection cover will be the power back up, which will get activated in fake switch off mode.

We have seen regularly every day the number of atrocities against women and girls are growing and are rising day by day.

Anyhow, every individual is carrying a smartphone nowadays and we can modify it to protect the individual; it will be a success story and we can grab the culprits instantly.

DESCRIPTION/BENEFITS/FEATURES/EXPLANATION

An app is developed for the safety and protection of the user.

It is called a fake switch-off app.

It is used for the protection of individuals and reporting instantly to alert the authorities and grab the criminals.

Nowadays, how easily the culprits get away even though the individuals are carrying mobile phone because they are unable to capture the proof, this app helps by automatically starting to work in secret mode and giving out the required information to the authorities.

Even the phone will be touch-sensitive and will be set as per the user how he handles the phone and will take and save the manner in which the user is using the phone; if the phone is left or forgot, then no one can touch the phone or switch off the phone, the actual switch off will be modified as per the user and will have a passcode or pin or a pattern.

Even the user can set the phone to allow it to take videos or record voices in secret mode when domestic violence is taking place and proof is needed for giving it to the authorities.

The power back up will be in the phone cover itself and cannot be separated easily.

The systems can be modified as to how to alert the people or ICE contacts the hooter, the siren and the strobe light and the video recording and voice recording and all the features will be user friendly and can be set by the user and the passcode and pin and pattern for the actual switch off.

Even the system and the software will be linked to the bank account and in any emergency situation, the transaction can also be faked and in the event of a forceful transaction, it will show all the procedure exactly the same but in reality, it will send out the information and location to the authorities and ICE contacts.

The login and account will have the same features and will have security features and will confuse the forceful transferors.

All the ICE contacts will be alerted in case someone is not available and cannot alert the authorities, and the mail will be sent along with the pics and photos and video recordings and all required information including location to the fullest information to grab the culprit.

The purpose of the app is to utilise the AI properly and in a useful manner and create jobs and doing R&D

and show what engineers are capable of and even can design smartphones at cheaper prices so that all the women and girls are safe in the country by purchasing these devices.

CHAPTER 18

MODERN SHOWER SYSTEM WITH AN INJECTION SYSTEM

FOAM, SHOWER GEL, SHAMPOO, CONDITIONER, FRAGRANCE, ROSEWATER AND FINALLY HOT AND COLD WATER.

WILL BE A HIT AMONG ALL AGE GROUPS, KIDS AND ELDERLY AND ESPECIALLY WOMEN/GIRLS.

ALL THESE WILL BE FROM A POWER SPRAY FROM COMPRESSOR FITTED OUTSIDE THE WASHROOM OR ANY CONVENIENT PLACE.

FOR ELDERLY WHO CANNOT APPLY SOAP AND SHAMPOO AND CANNOT TAKE A SHOWER EASILY, THERE WILL BE A JET SPRAY AT KNEE LEVEL AND ONE JET SPRAY OVERHEAD ADJUSTABLE UP AND DOWN AND SIDEWAYS.

INTRODUCTION/PURPOSE

It's time to improvise the showering experience and make it easy for all age groups and have a healthy and hygienic bath just by relaxing and standing, and we have been using for ages the same conventional method of shower and bathing.

It's stress-free bathing for kids and elderly.

A modern shower with an injection system includes foam initially, then water and shower gel and shampoo and conditioner and fragrance and rose water and so on and can be selected with the help of a waterproof remote.

The containers and the reservoirs will be connected and will operate as per the requirement or command from the input remote and even there will be a screen, which will show all the necessary data, the temp of the water and the fluid flowing and percentages left upon selection of the options in the remote.

The reservoir boxes will be readily available from the market.

The compressor will give power spray for water and fluids.

The standard box sizes for the reservoir will be standardised by various brands, which contain all the fluids necessary.

The injector will have dispense boxes, which, along with the small compressor, will serve as per the selection.

FEATURES/BENEFITS/EXPLANATION/DESCRIPTION

The main purpose of this innovation is to give a stress-free bathing experience and a hygienic shower and refreshful effect.

We have seen kids and elderly suffering from having a neat and clean bath, they cannot apply soap or shampoo and cannot sit or stand; so to make it easy for them, the modern shower is introduced.

The modern shower is with the compressor and injection system to cater to all needed fluids and gels.

The reservoir will be like a water heater or can be integrated with the water heater, wherein all the boxes carrying fluids can be dispensed at the command of the remote input and digital display.

The pressure of the water will be well within limits that the body can withstand and there will be a child lock for the temp of the water so peacefully that even kids will and can take shower even if the water heater is on and the sensor won't allow the temp of the water to harm the skin.

The sequential bath will be more effectively performed with the help of pressurised blast of the fluids and even water and people will get good sleep after a long day's work.

The sequence makes it easier for all age groups and just stand there, enjoy and feel fresh and active.

There will be two outlets, one at the knee level and one over the head for ease of washing the lower body parts, which will be adjustable side by side and up and down as per the height.

Even an elderly sitting on a waterproof chair can just sit and relax and play with the remote and finish the shower and get a hygienic wash.

The liquids and fluid boxes can be purchased from various brands and are available to fit the reservoir.

The digital display of the injection system will be in sync with the CCTV of the home and can only watch the main door if any stranger arrives and can interact with any delivery persons and visitors.

CHAPTER 19

PEDALLING CYCLE FOR POWER GENERATION

PEDALLING CYCLE FOR POWER GENERATION AND THE POWER IS STORED AND CAN BE UTILISED FOR THE HOUSEHOLD CONSUMPTION LIKE WATER HEATER, OVERNIGHT BEDLIGHT, FAN, MIXER, ELECTRIC COIL OR INDUCTION-TYPE COOKER.

THE POWER WILL HAVE AC TO DC AND DC TO AC OPTIONS AS WELL.

INTRODUCTION/PURPOSE

The pedalling cycle modified for maintaining health and regular exercise will generate power as well, which could be stored and consumed and can save the utility bills.

Every individual of the household can perform the exercise by pedalling and in return, the power can be utilised for household electrical power consumption.

The pedalling cycle can be pedalled and the power generated can be stored and can be utilised later without wasting the government power source and even has health benefits.

It will be portable and can be moved from one room to another and can be used for household equipment like electric cooker, water heater, washing machine, fan or AC and fridge as well.

BENEFITS/PURPOSE/EXPLANATION/FEATURES

The innovative pedalling cycle will maintain health as well as the power for utilising the household electrical equipment.

A lot of people spend unnecessary amounts in gyms and still can't get the required results, but this innovation gives us both health benefits and power for domestic utilisation.

The pedalling will be both with my arms and legs as well, the exercises for both upper and lower body.

The power generated will be showing on a digital screen and the equipment will have male-female sockets for ease of use and all the members of the household can contribute to the power saving and power generation.

The pedalling cycle can be modified to look like a sports bike or car and can be even pedalled with hands and legs and early in the morning can make a habit of doing whole-body exercise and still getting paid by saving power.

Chapter 20

APPS FOR ENDORSING LOCAL-MADE PRODUCTS

APPS FOR ENDORSING LOCAL-MADE PRODUCTS AND SUPPORTING ONE'S COUNTRY AND DEMOLISHING THE VIP AND ELITE CLUB AND CELEBRITY CULTURE

INTRODUCTION/PURPOSE

The app is developed to support the local-made products and abolishing foreign products.

Earlier in the 100 professionals chapter, I discussed supporting local-made products and bringing down the price of the products by not giving the ads to the celebrities where the price of the product goes up.

The details will be shown in the app how the locals are dependent upon the local and organic products and how good is to use them rather than blindly spending lots of money on foreign products.

Even the celebrity culture should be abolished and show the reality to the people how the celebrities are paid lots of money and in reality, they never use the product.

DESCRIPTION/BENEFITS/EXPLANATION/FEATURES

The main purpose of the app is to support local-made products and demolish celebrity culture and knowing the true benefits of the products compared to foreign products.

The app will show all the local products and their benefits and needs no advertisement to support one's country.

All the items will be covered in all the categories and will show in detail all the benefits and how to strengthen our economy.

We should abolish the culture of flashing the foreign products and items and feel proud about it and feeling one is above all by purchasing the foreign brand products.

We should support our country and support our people and abolish foreign products and even food chains as well.

The app will show all the information like that of Wikipedia for all the local products and in all categories and all the required information.

Creating awareness will give us the benefits and strengthen the economy and make the country stronger.

CHAPTER 21
LOCAL TOURISM HOLIDAYING

LOCAL TOURISM HOLIDAYING AND SITE SEEING THROUGHOUT THE COUNTRY IN A VANITY VAN LIKE THAT OF CELEBRITIES WITH ALL THE FACILITIES AVAILABLE IN THE VAN

BETTER THAN WORLD TOUR BECAUSE INDIA IS SUCH A BEAUTIFUL COUNTRY AND EXPLORING INDIA WILL REVEAL MANY SECRETS AND GIVE US INNER SATISFACTION BY GETTING TO KNOW THE DIFFERENT PEOPLE, DIFFERENT CULTURE AND VISITING DIFFERENT WEATHER CONDITIONS AND TASTING DIFFERENT PLACES FOOD AND SPECIALITIES.

ONCE IN A YEAR, HOLIDAYING WITH FAMILY WITHOUT HOTEL STAY WHEREVER TRAVELLED IN A VANITY VAN WITH ALL THE FACILITIES AND FEATURES AND COMFORTS AND STILL FEELING LIKE HOME.

INTRODUCTION/PURPOSE

An app is developed by giving details of a holiday package or else a surprise visit is decided by the company to take people to a surprise location, undisclosed location as per their budget.

As per the requirement of the customers, a journey and a holiday will be planned with all the family members and a vanity van to fit them all will set out to the holiday place.

The team will have cooks and drivers and stewards and stewardesses and there will be no hotel stay at all even if the journey is for weeks.

Any camping can be done at any place they feel is safe or else will have enough bedrooms within the vanity van, overall a new experience.

The kitchen will be active 24 hours and serve them fresh hot food and whatever they need.

DESCRIPTION/BENEFITS/FEATURES/ EXPLANATION

India is such a beautiful country and we need to explore it more often to get the real taste and essence of it that too in road travel and camping in the fields and forests and villages.

An app will show all the places to be visited but the names will not be disclosed just by looking at the natural beauty, they will decide that they need to cover all the places in one trip and can set out on the journey and book the caravan and vanity van.

The main purpose is to come closer to our country and no country can match the scenic beauty of our country, our heritage, greatness, harmony, unity and essence.

A holiday with family across India is worth a memory of a lifetime and coming close to our country and our people as well.

No journey to any country can match this journey

Travel experts know all the beautiful locations and across north, east, west and south and will give all the locations and required information as per the requirement of the customer and can set out on a journey.

All the necessary precautions will be taken before setting off, including a security guard for the safety of the travellers.

No hotel stay will be there; only camping near rivers, mountains, fields and villages bring more joy to the journey; it is a kind of adventurous travel and a memory for a lifetime.

The kitchen facility along with the travel gives more pleasure and even the w/c and shower inside the vanity van gives a heavenly feeling, without even stopping the vehicle for relieving.

The main purpose is to bring the 7-star facilities and prove that even a common man can travel like the rich, privileged and celebrities and enjoy life and the comforts of a super master class vanity van travel.

What is the point in going on a foreign trip when we have everything here and can take all the members of the family and get closer to one another and eliminate the differences, forming a bond?

All the travellers who travel across the globe will say that India is the country worth watching and with different cultures and different languages, still, a unity exists in India.

Vanity van journey enjoying road travel, holidaying trekking, camping cam fire, cooking, photoshoot recording all the journey once in a lifetime must do it and add to the bucket list for every Indian for the sake of their family.

Chapter 22

HOTEL/RESTAURANT ALL OVER THE CITY

YEARLY CARD AND ONLY COST TO COST, NO PROFIT

VEG/NON-VEG CUISINE

GOVERNMENT CENTRES NOT PRACTICAL

PRIVATE HOTELS/RESTAURANTS WITH COST TO COST NONPROFIT BUT WITH YEARLY CARD FACILITY ONLY

PURPOSE/INTRODUCTION

Restaurant food is expensive and is not affordable by all, so regular people and common people moving around the city need some arrangement of breakfast, lunch and dinner while taking care of their jobs, either they have to go for roadside food or carry a box packed from home or else find a cheap place where they can feed themselves and this is a daily routine.

This concept solves the problem, food places should be opened all over the city as a non-profit organisation and will serve three times food at cost to cost basis both veg and non-veg, but the food will be served to those customers paying for a yearly card for e.g. if daily meals come to around 100 rupees at the cheapest then for the yearly card, it comes to around 36,500 rupees, but this option and concept will get the card only for 25,000, which is the best bargain, and one can get the homely food and hygienic food all over the city and no need to carry tiffin and lunch boxes and unnecessarily spend more money on restaurants and one will have a nice hygienic place to peacefully eat and relax and doesn't have to worry about the expenses if a person normally eats outside every day when it comes to around 200 per day, which includes the profit for the food place which they charge.

This concept will help all categories of people and help them financially to save money and even jobs are created in the hotel field and nobody will be jobless and will earn their own living.

If a person is paying 25,000 per year, then per day it is coming to around 70 rupees, which is even cheaper than the home food expenses, or the concept can be modified for only two meals a day and the charges for yearly can be brought down as a lot of people have breakfast from home and may not need the breakfast option and some people may need breakfast but by dinner, they will reach home so whatever the case, three times a day yearly package and two times a day yearly package will be available, thus helping only the citizens.

The business will cover all the expenses and even the citizens will get fresh hygienic and homely food three times a day.

If 100 centres are opened around the city, then all the city will be covered and even the overheads will be covered for the salaries of the hotel staff.

If around one million people are fed every day and they have paid yearly card, then the amount comes to around 2500 crores and it is a huge amount and needs a lot of planning and staff to maintain and feed all the people around the city and look after the rent and salaries and overheads and in turn, everybody is benefitted. The jobs are created in all fields and everybody grows simultaneously.

Just imagine if in one city the amount generated is 2500 crores yearly, then in about 100 cities how much will be generated 2,50,000 crores and imagine the job creations and the system to maintain all this, how huge the setup will be from purchasing to cooking to cleaning, etc.

The main purpose is to benefit all.

We have seen everyday restaurants closing due to losses and this concept will be evergreen and if all the expenses are covered, then the profit will be utilised to open more branches to accommodate whole of the city.

Even the people who want to donate generously can sponsor for one month and without displaying or disclosing their names can feed people and support the city.

A committee can be set up to maintain the hotels and keep the accounts transparent and clear.

An app will show the locations of all the food places and help the commuters and even the concept can be expanded to highways where the food is very expensive and the places where the buses private and government stop frequently and facilitate the travellers.

Chapter 23

CAREER GUIDANCE/ INVESTMENT GUIDANCE/ LOSS PREVENTION/ ADVISORY COMMITTEE GIVES SERVICE AT SOME CHARGE

SAME AS CHAPTER 17, EXTENSION OF THE SAME CONCEPT

PURPOSE/INTRODUCTION

A lot of people need assistance and guidance from experts regarding education, studying in foreign universities, career, investments and mental health/ depression and anxiety issues.

People want someone to hear them and understand and give valuable advice.

A lot of problems come up only because of the financial losses so before investing huge amounts, one can have a healthy chat with the experts and do prevent the loss and get advice from the market experts and people who have studied the market and explain the risks involved to them and convince them to save money and slowly invest small amounts with good benefits instead of overshooting the suggestions and falling prey to the market risks and then repent.

An app will help in getting proper guidance and at a charge will visit the homes as well.

Even can provide a counselling session for the improvisation of relations and domestic violence issues and parenting issues and understanding the present trend and how to treat the elders and kids at home and all sorts of issues and helping out all age groups to overcome their problems and lead a happy life.

Explanation/details/features/description/benefits

An app will help in giving free suggestions and at a charge will give detail counselling and solve the problems.

Any issues related to anybody regarding anything will be listened to and expert advice and schedule will be given to follow to succeed and make them comfortable and making them aware of the importance in the family and creating awareness of one's self and the talents one has.

If any other advice is needed, then they can have it from outside as well from famous professional people.

The main purpose of the concept is to bring down the suicidal tendencies in any individual who is disturbed in life because of their issues and give them hope to live and show them how they can be a better and responsible human and help others and lead a happy life and overcome all sorts of fears and shortcomings and problems and get that stamina to fight the odds and hope for a better tomorrow because any individual who commits suicide will take all his family into depression and the loss can never be fulfiled or overcome so that before anything happens, we have to act and release all the negativity and bring positivity into their lives.

Finally, the team will be motivational speakers and even religious speakers on demand can be provided and get the family going and teach them to get most of the life out of every moment.

Chapter 24

BED MODIFIED WHICH CAN ACCOMMODATE A/C, HEATER, FAN AS PER THE WEATHER CONDITIONS FOR GETTING COMFORT AIR TEMPERATURE

DC OPERATED AND CHARGING FACILITY WITH AC CURRENT.

ALL INSULATED AND WITH SPECIAL MATERIAL TO AVOID ANY SHOCKS AND STATIC ELECTRICITY.

THE USERS OF THE BED NEED NOT USE ANY FAN OR A/C OR HEATER DEPENDING ON THE WEATHER CONDITIONS AND HAVE A PEACEFUL SLEEP.

THE CONCEPT IS TO GIVE ALL CLASSES OF SOCIETY A PEACEFUL SLEEP AFTER A HARD DAY'S WORK AND DON'T FEEL THE PRESSURE OF NOT OWNING A/C OR HEATER OR FAN.

THE BED WILL BE AVAILABLE AT A PRICE VERY LESS COMPARITIVELY TO OWNING AN AC FAN OR HEATER FOR GIVING US THE COMFORT

AIR TEMPERATURE AND STILL BE HYGIENIC AND PROPERLY SANITISED AND CLEAN, WILL HAVE VERY LESS MAINTENANCE COST, AND CAN BE DONE AT HOME ONLY, NO NEED OF ANY EXPERTS OR TECHNICIANS.

INTRODUCTION/PURPOSE

A bed is modified in such a way that an AC/heater/fan is inbuilt for experiencing different weather conditions, available at an affordable price even for the economically weaker sections of the society, to have a king's sleep.

Normally, a room split or window AC is not affordable for everyone and even the load of power bills can't be taken, and while using these, all the room is cooled unnecessarily whereas each person requires only very less cooling to get a comfortable sleep.

As we know cold air is heavy and hot air is light so all the hot air will concentrate in the upper part of the room and the bed will be cold and give comfort air temperature within seconds, so is the concept of a bed with an inbuilt AC/fan/heater at an affordable price with the HEPA filter and blower fan and air terminals adjustable grilles and charging during the day and consuming the power during the night for running the AC.

BENEFITS/PURPOSE/FEATURES/ DESCRIPTION

Even the middle class and poor have the right to get comfortable air temperature in summer/winter and under extreme weather conditions.

So the innovation is made wherein the bed will have AC/heater/fan and will create a comfort air around the bed and accommodate the sleepers on the bed and a special mosquito net can be hanged from the ceiling and can be adjusted left or right, up or down and cover the bed and be locked with the bed.

Normally, we see most of the time that the refrigerator will be running unnecessarily with less load and there will be an option with the copper piping to utilise the refrigerant from the fridge and utilise the cooling effect and transfer it to the bed AC.

There will be an option of connecting it from the fridge by using the copper pipes and make use of the refrigeration cycle and achieve a cooling effect, the refrigeration liquid and the system can be used to get the chilled water or gas or fluid to flow through the cooling coils of the bed and with the blower running will supply the cold air and the speed of the blower can be regulated as per the requirement and the swing in the terminal grilles is adjusted to get the proper effect, giving the effect of fresh cold air rather than cooling the whole room unnecessarily by wasting the power and energy.

The air circulation will be done with the help of an air transfer system and the HEPA filter to control the bacterial infection and dust prevention.

The purpose of this innovation is to give the poor and middle class the comfort of the AC and heater as per the weather conditions.

This innovation will be silent enough to match the noise criteria and will not disturb the sleepers and will have additional benefits of power and energy-saving unlike that of a split and window AC.

This can be modified to use the refrigerant from the fridge or else use only chilled water to get cold air by adding the pumping and circulating technology.

The DC will be used to run the bed AC and with special bed and material will give nice comfort air and noiseless AC and will be shockproof and all the materials will be insulator material, which will resist the current flow.

There will be a touch screen to adjust all the features or else a remote.

A proper circulation and drain facility for condensation issues will be designed, even there will be bio mode to adjust as per the existing weather conditions.

The bed gives cooling and heating atmosphere around the bed unlike the A/C room split or window and unnecessarily cools the room and creates noise.

This will be a silent AC with both AC and DC for charging and discharging.

All the features of a room AC will be present in this innovation.

The grilles and air throw can be modified as per the orientation of the room size and requirement of the customer.

Instead of purchasing a separate bed, separate cooler, separate AC, separate fan and separate indoor and outdoor units, this innovative bed will be affordable by all classes of people and one can enjoy all the comforts of life, save money and have a royal life.

The mattress along with bed will be a respiratory type and anti-fungal and anti-bacterial one and can be washable and with the orthopaedic concept.

This innovation is a masterpiece withholding cold airflow, filtered airflow less power consumption silent one all washable and easy maintenance one with mosquito self-net suspended from the ceiling covering the whole bed and net can be moved and adjusted left, right and up, down.

This innovation will easily adjust to the gated communities where there is a chilled water facility from the district cooling system.

Or else an apartment can place a chiller on the roof and every home can be given chilled water outlet and inlet and can use this bed by connecting with proper valves and meters insulation and cutoff facilities.

Even the refrigerant vrv/vrf concept can be applied to the apartment and rented out with a central refrigerant reservoir on the roof of the building, supplying the refrigerant to all of the flats for monthly rent, and we can utilise the refrigerants for fridges, AC and beds and can be free of the maintenance burden.

The system will help in eliminating the individual burden of buying and spending money on power bills and maintenance.

The system who understand the terminology of FCU chillers and vrf and vrv will get to know about this innovation and know the concept of how it can be modified to improvise our living and with less investment and more savings.

Chapter 25

PROTECTION OF INDIAN CITIZENS AROUND THE WORLD

CREATING AN APP IN WHICH ALL THE TRAVELLERS WILL GET REGISTERED AND WHO ARE ALREADY STAYING IN FOREIGN COUNTRIES AS WELL WHETHER FOR THE JOB OR BUSINESS OR PLEASURE FOR MEDICAL TREATMENTS.

THE PURPOSE OF THE APP IS TO HELP OUR FELLOW INDIANS WHEN IN DISTRESS AND FORWARD THEIR CONCERNS TO THE AUTHORITIES AND HELP THEM AT THE EARLIEST TO COME OUT OF THE CLUTCHES OF THE CHEATERS AND HUMAN RIGHTS VIOLATORS AND SAVING THEIR KIDS AND FAMILY AND KEEP A CHECK ON THE GOVERNMENTS PERFORMANCE AND THE APP WILL SHOW THE SERIES OF PLAN OF ACTION AS TO HOW THEY WILL BE RESCUED FROM THAT COUNTRY AT THE EARLIEST.

NEARLY THOUSANDS OF INDIANS TRAVEL ABROAD TO GET LIVELIHOOD AND ARE CHEATED ON A REGULAR BASIS AND CAN'T COME OUT OF THE SITUATION AND ARE

EVEN PUT IN JAILS FOR NO REASON, SO IF WE HAVE INFORMATION OF THE PEOPLE, WE CAN IMMEDIATELY HELP OUR FELLOW BROTHERS AND SISTERS TO MAKE USE OF THE FACILITIES PROVIDED BY THE INDIAN AUTHORITIES IN THAT PART OF THE WORLD.

THEIR CASE CAN BE FORWARDED TO THE CONCERNED AUTHORITIES AND CAN BE MONITORED ON A REGULAR BASIS AND UNTIL THEY ARE OUT OF THE SITUATION, IT'S THE RESPONSIBILITY OF THE PEOPLE BACK HOME TO LOOK AFTER EACH OTHER EVEN WE ARE FAR; THAT SORT OF ATMOSPHERE NEEDS TO BE CREATED AND PUT FEAR IN THE MINDS OF THE HUMAN RIGHTS VIOLATORS, NO MATTER WHICH COUNTRY THEY BELONG TO BECAUSE IN THE FUTURE, INDIA AND INDIANS WILL BE RESPECTED FOR THEIR INNOVATIONS AND TECHNOLOGICAL ADVANCEMENTS AND ACHIEVEMENTS AND WILL NOT NEED TO GO TO ANY COUNTRY TO GET LIVELIHOOD; INSTEAD, EVERY OTHER COUNTRY PEOPLE WILL LIKE TO COME TO INDIA FOR LIVELIHOOD AND WORK HERE AND LEARN FROM US AND COME ON A WORK VISA.

INTRODUCTION/PURPOSE

An app created for the register of Indians travelling abroad if alone or with family, their permanent address and what's the purpose and which country they are travelling to and how they can get help in any situation or else any help soon after landing, all the details and the volunteers willing to help will also put out their contact details and will save from the cheaters.

Regularly, there will be information sharing in any form, may be any help or favour or else even in the financial angle, etc.

BENEFITS/FEATURES/EXPLANATION/DETAILS

In the past few years, we have seen our Indians being targeted abroad especially in the Gulf how they are treated and even put up in jails because of no reason, they don't have access to Indian authorities and even are foodless for days and have to spend the nights on roads and the sponsor who called them is enjoying and is least bothered about their lives.

Even someone willing to do charity can help these people and can be brought back safely and even we can have a check on the system and authorities, how they are helping our people abroad and saving them from the clutches of the human rights violators and all this information will be in the app and there will be a group and anybody can access it from anywhere about any country with the Indian population in that place.

Someone somewhere is suffering and we don't even bother about our fellow Indians is not good for the society; it can happen to our near and dear so creating a group and keeping a check on their well-being will bring out the peace and harmony back in the home country, and a lot of people willing to help will come forward financially and even solve their problems back home with their dependents.

In the near future, India is going to rule the world and we should know how the countries are treating our people so that in the future when they come to our place, we will show our values as to how one should treat another human being and respect.

People will stand in lines to get a visa to come to India and will be getting rejected if not able to show a proper reason.

People will crave the citizenship of India and Indian companies will hire foreigners for getting their jobs done and we will sit and enjoy the benefits, people will come from all over the world to learn the new technologies and will be willing to settle here and get citizenship.

The next decade is going to be crucial for the development of the country and will come out with flying colours and we will rule the world and so many intellectuals will come out and will do some good for the economic strengthening and development and will come out with newest software and apps and cutting edge technologies and will sell it to the rest of the world.

Never underestimate the capability and capacity of our Indians and our country, the message is already going out and will go out even more strongly in the near future.

Finally, we should make use of the resources and try to bring out the best in us to find a solution to the problem and observe, grasp, analyse and give the best solution to the betterment of the society.

We Indians are not concentrating but a time will come when every individual will be capable of giving solutions to the problems of the world.

The main criterion is to look around, observe, grasp, analyse and apply our thoughts in giving solutions to the problems around us and make use of the intellect and meditate, imagine, explore and travel across the universe through concentration and meditation and take all the positive energy and imagining things, which people have never imagined and try to create and pen it down and counter it with questions and take out a solution and create a world of our own and imagine new apps and possibilities and innovations and do research and apply one technology to assist some other technology and integrate and imagine the future even when we are gone and how the world will develop in the future and what are the possibilities and how the technology will develop and how we can bring the 100 years ahead technology in today's time and keep on concentrating and how we were 20 years back and how we will be 20 years from now in the future, will bring out the best in you and will lead to wonderful results, thus making everyone an innovator and a creator and visionary and in the end, we can see the future with much clarity and imagine what it's going to be like, we have the ability to bring something from the future and apply in today's time.

Finally, do not limit ourselves to the existing jobs but to make use of the intellect to the fullest and imagine and think like a master and a creator and the king/queen of the world and imagine the life on this planet 100 years from now and what are the possibilities and how we have evolved over the centuries and how we keep on evolving and what best we can do to protect the humans from any disasters and natural calamities and how we give solutions, all these matter and will make a human a superpower and the brain will grow and expand like the universe, we will not be here but our thoughts will be there forever with the next generation and they will discuss our ideas and discuss our achievements and take the concern to the next level and will find a solution and be prepared for any situations.

Chapter 26

PROTECTION SUITS FOR TWO-WHEELER RIDERS AND PASSENGERS, NON-AC CAR DRIVERS AND PASSENGERS

PROTECTION AGAINST SUN HEAT, DUST, RAIN AND VIRUSES, BAD WEATHER AND WITH REFLECTORS.

INSIDE THE SUIT, THERE WILL BE AIR CIRCULATION WHICH WILL PROVIDE WITH THE COMFORT AIR TEMPERATURE AND PROPER VENTILATION FOR THE GASES TO ESCAPE

INTRODUCTION/PURPOSE

In today's world, where we are daily prone to different contagious diseases and viruses, there is a special suit designed to protect from all these and even gives comfort air temperature and with ventilation and it can be washable and easily foldable and placed in the boot space.

We have seen how people travelling to outer space wear protective suits and helmets, likewise but at a cheaper rate with all the features same to give comfort to the people and safeguard them from all viruses and weather conditions and feel fresh wherever they go, this suit will be designed.

The suit will be having a DC power source to give a cooling effect or heating effect depending upon the weather condition and can be charged at home; it will cover from head to toe and anybody can wear this easily and protect themselves from viruses and keep themselves safe.

Normally, people travelling on two-wheelers or non-AC cars will be tired of extreme heat and this suit gives them comfort air temperature and keeps their vitals working normally and this suit even protects from rain and viruses and extreme weather conditions like storms, dust and pollution.

EXPLANATION/PURPOSE/FEATURES/DESCRIPTION

To overcome the extreme weather conditions and safeguard ourselves, the suit is designed at an affordable price, which keeps the people safe from heat, dust, rain, storm and viruses.

Not only in pandemics but we need these suits during normal days to protect from pollution and extreme weather conditions.

The two-wheeler riders are the ones who are the most tired of the sun heat, rain, dust and pollution and even the people who travel in buses and non-AC cars.

The purpose of the suit is to give fresh and healthy air and keep away the viruses and it will have a DC power and will have air-conditioned air and will maintain comfort air temperature in the suit.

The suit will maintain the health of the people who will always feel fresh by wearing while travelling.

Even people going to far off places to attend the functions can wear this and travel in the hot sun; this will not affect them.

The suit will be designed and available to the public at affordable prices and will be a beck mark in the industry and keep the citizens healthy and active and can perform better.

The sizes will be for kids also.

Poor middle-class families travel on two-wheelers and still will be protected from the sun, rain, dust and viruses.

Even the clothes are not soiled and one can attend interviews fresh like a flower.

In the near future, we need these types of suits to keep ourselves and our kids safe from dangerous viruses and hence this needs to be introduced in the market at the earliest.

After analysing the oxygen levels in the air, the pollution is increasing every day and because of this, we are prone to diseases; we have to start taking the issue seriously and millions of people are losing their lives because of air-borne viruses and contagious diseases.

Chapter 27

FLYING VEHICLE RAPIPORT

COMMERCIAL TO CARRY PASSENGERS AND PERSONAL VEHICLE

ALL THE TRANSPORT VEHICLE COMPANIES LIKE THAT OF TWO-WHEELERS, FOUR-WHEELERS, SIX-WHEELERS OR ANY AUTOMOBILE COMPANIES IN INDIA SHOULD START THE MANUFACTURING OF THE FLYING VEHICLE, MAKING USE OF THEIR MACHINERY TECHNOLOGY AND R&D DEPARTMENT AND LAUNCHING THEIR MODEL OF FLYING VEHICLE AT THE EARLIEST.

INSTEAD OF RUSHING WHEN DEMAND IS HIGH AND BUYING IT FROM THE FOREIGN COUNTRIES, RIGHT NOW THE WOLRD IS EXPECTING TO TRAVEL THROUGH AIR AND REACH THEIR DESTINATIONS FROM HOME TO OFFICE OR HOME TO BUSINESS PLACE AND HOME TO SHOPPING, HOME TO SCHOOL; BUT THE CRITERIA IS THE VEHICLE SHOULD BE IN OUR GARAGE OR ON ROOF AND READILY GET THE ACCESSIBILITY TO FLY ANYWHERE FROM ANYWHERE AND STILL HAVE THE SAFETY FEATURES SAME AS THAT OF A COMMERCIAL FLYING MACHINE.

ALL THE COMPANIES ARE HAVING ALL THE DEPARTMENTS WITHIN THEIR PLANTS, NO NEED TO BUY ANYTHING SEPARATE, JUST NEED TO PUT SOME EXTRA EFFORT AND LAUNCH THEIR MODEL AND LEAVE IT FOR TESTING AND EVEN IN TODAY'S WORLD, ROBOTICS AND ARTIFICIAL INTELLIGENCE ARE GAINING SPEED SO NO NEED TO LOSE ANYTHING. ALL HAS BECOME SO EASY. WE JUST NEED THAT CONFIDENCE AND PLANNING AND ORGANISED STRUCTURE TO SHOW THE WORLD THE CAPACITY AND CAPABILITY OF THE INDIAN COMPANIES.

ONLY ONE COMPANY IS DOING THE R&D IN THIS FIELD THAT TOO FROM SUPPORT FROM THE GOVERNMENT, WHEREAS BY LOOKING AT THE PRESENT SCENARIO, IT WON'T TAKE LONG TO BRING OUT THE BEST IF THE PRIVATE COMPANIES DO SO AND BRING OUT THEIR PROTOTYPE AND START TESTING MAY BE A HOVERCRAFT OR FLYING BIKE OR FLYING CAR AND BE THE TOP IN THE FIELD.

WE HAVE LOT OF AERONAUTICAL ENGINEERS AND WHEN WE CLUB TOGETHER THE AERONAUTICS AND MECHANICAL AND AUTOMOBILE AND ROBOTIC ENGINEERS TOGETHER, THEN IT'S THE BEST TEAM WHO CAN DO WONDERS AND BE AT THE TOP SLOT IN THE FIELD.

INTRODUCTION/PURPOSE

The flying vehicle concept is not far from reach if the Indian companies have a plan and target, and it should be done at the earliest and even the companies making drones can also bring out the prototype; this shows the ability and the best engineering technologies we can produce.

What's the point in rushing when the demand is high and copying the foreign countries, right now we have all the infrastructure, technology, manpower, machinery, robotics and best electrical and mechanical students and engineers and investing in the concept will bring out the jobs and even there will a time when we can start exporting our flying cars to the rest of the world.

BENEFITS/FEATURES/EXPLANATION/ DESCRIPTION

Right now the whole world is in the process of creating and producing their own models of the flying vehicle concept for personal and commercial use, which can readily have the accessibility to take off and land from our backyard and the manufacturing is also started and experiments are going on at a very fast pace, and there is a requirement to show the world our masterpiece at the earliest.

There are a lot of passionate and visionary engineers who can do so in six months but still, we don't see any reaction from the big companies and encourage talent.

We have the capability of delivering the world's cheapest flying vehicle with all the technologies fitted and the safest model as we have seen in the concept of sending satellites to mars (108) one after the other without any fail, snag or glitch.

Around the world, we see how millions of dollars are being invested and the R&D behind the concept is taking place and the testing is going on propeller type, vertical take-off type, turbine engine-powered rotor blade type and high-speed fan-powered type of flying vehicles.

The concept is very easy when all the engineers are combined and make a dream team and can deliver it to the world we have a lot of engineers who are experts in propellers, turbine engines, rotorcrafts, wind tunnels, CFD analysis, aerodynamics, hydraulic systems, pneumatic systems, airframe and structure landing and take-off gear, but putting all together we can deliver a masterpiece and even can export it to the world and become the world leader.

Chapter 28

MOVIES 4D CONCEPT WITH SCREENS

MOVIES 4D CONCEPT WITH SCREENS ALL AROUND WITH FOUR VIEWS AROUND US AS OF THE ORIGINAL PLACE, THEATERS AND SCREENS TO BE MODIFIED AND ENDING THE AGE-OLD 2D THEATER CONCEPT

INTRODUCTION/PURPOSE

Introduction of the 4D theatres' concept and 4D films and even the theatre screens to be modified.

We are capable of seeing around us 360 degrees with a little help of turning and tilting and head movements; likewise, when watching a movie the theatre screens should also be all around us and we should be able to watch the real world, not just in 2d screen in front of us but the real world.

If a person visits a place and wants us to enjoy the beauty of that place, then he takes video of that place and sends it to us and we can see only one view at a time so introducing the concept of able to see all the views as he sees it and enjoys the scene, a four-view camera/video recorder is created and can capture the four-side views and transmit it to the recipients and the receiver will be able to view all the views in the screens all around him as the same as that place, the front view will be displayed in the front screen and left and right views in the respective screens and the view behind will be shown in the screen behind us, thus creating the same scenario as if we are present at that place and watching the scene, the same is the concept with the films making all the sides should be covered and displayed for the viewing.

The main concept is to film the four sides of that place and display it to the recipient and the receiver should be able to see it as if he is really present and watching the view.

The original video and the sent video will have all the same four sides recorded and captured and there will be no difference between the original place and the video recorded.

This type of camera/video recorder is created and the video can be shoot and sent to our near and dear, the camera can be worn like a helmet or else can be worn like a vest and will be able to record all the four sides and transmit it in the same way thus giving the same effect as that of the original shooting place.

For viewing this, a screen like that of ISLEEP will help to record four sides and view at the same time; even live transmitting can be possible, which is why there is a need to develop the 4D theatres and screens.

PURPPOSE/BENEFITS/EXPLANATION/DETAILS

Ending the 2d theatres' screen TVs concept and bringing the virtual reality of 4D concept by covering all the four directions and displaying in the screens around us and enjoying the views as that of the original place.

Even the films can be made in this way and it will be the future concept, so the real world should be shown not like in the 2d screens where only the front of the camera is displayed.

This is one type of the VR concept.

The main criterion is to create a camera and video recorder so powerful and one can capture and transmit the same way as that of the surroundings and the transmitted image should be properly aligned and displayed with the overlapping of all the sides and giving proper image and video.

The video camera can be popped up in the smartphone and should be able to shoot the four sides and properly align and transmit it to the viewers on their screens.

The ISLEEP concept will be best to view the videos, even the movie screens can be modified to watch the real world and even movies should be taken in the 4D mode and covering all the sides and imagining as if we are in that movie.

Even the roaming concept as explained earlier matches with this video recording and gets the real world transmitted to the bedroom.

Even the videos from space can be taken in four views and transmitted and more highly sophisticated videos will be to mix with the real effects of air, speed, temperature, odour and all the real effects and the feeling of movement as well.

Al the ambience and effects will be added and transmitted to the customer as per the request, originally the real video and later when modified will get all the real-time effects like heat, air, speed, temperature and smell and odour of that place.

The videos taken will be modified and the viewer can feel the real effects as that of the original place, for e.g., from the desert, from the sea and from cold mountains all these effects will be taken from the original place and added and modified and transmitted to the customer to get the real effects and the ISLEEP CONCEPT is a suitable platform for watching all these and enjoying.

Even the movements can be felt in the ISLEEP from the simulation effect and with modification.

The movies will be made in this concept with all the parameters taken into consideration of movements with gyro and rolling, pitching, yawing and all the real effects.

Mainly the concept is to view the videos with real effects of the original place where the video is being shot, a raw video can be transmitted live or else to have a maximum effect can be modified and sent it to the customer as per the request.

Even the smartphone video viewing will be changed accordingly with all the four views being displayed on the same screen as front-back and left-right views.

The recognition of the artificial horizon comes from the gyro concept.

The concept used is not only for entertainment but to record real ambient conditions and transmit the same for the viewer from any part of the world to any part of the world and will include the original parameters like air, temperature, speed, odour and all other conditions, for e.g., snow, sun, rain, etc.

Finally, we should be the first to imagine and create and supply it to the world and become the master, so innovative thoughts are not to be taken easily and whatever we feel like we should make it to a real concept and apply and give the world a new technology every day.

Chapter 29

WATER BOTTLES DESIGNED FOR DISPLAYING

WATER BOTTLES DESIGNED FOR DISPLAYING THE QUALITY OF WATER, LEVEL INDICATOR, DISPLAY OF ADDED MINERALS AND METALS DISSOLVED, pH, QUALITY, WHETHER CONSUMABLE OR NOT, DISPLAY AND IDENTIFYING ALL THE DRINKS EVEN IF FILLED WITH SOFT AND HARD DRINKS AND DISPLAY OF ANY POISONOUS SUBSTANCES MIXED WITHOUT THE KNOWLEDGE AND CAP LOCKED WITH PIN FOR INDIVIDUAL USE AND SECURITY.

MAINLY FOR OFFICES AND HOMES AND EVEN OUTINGS LIKE IN A PICNIC OR A HOLIDAY.

THE DISPLAY WILL SHOW WHETHER THE WATER IS CONSUMABLE OR NOT AND WILL STORE THE DATA OF ALL THE LIQUIDS FILLED FOR THE LAST FEW DAYS AND EVEN GIVE THE INFORMATION OF THE CLEANING DATE OF THE BOTTLE AND FINALLY HELP US IN CONSUMING SAFE AND HYGIENIC WATER AND STAY SAFE FROM DISEASES.

EVEN THE TEMPERATURE OF THE LIQUID PRESENT INSIDE THE BOTTLE WILL BE DISPLAYED.

INTRODUCTION/PURPOSE

This innovative water bottle will display the required information, level, dissolved minerals, metals and substances, temperature, any bacteria present or fungus formation; it displays whether the water is consumable or not in the form of red and green display and the small display will be waterproof and along with this, the cap will be a pin code or passcode locked for the safety of the user so that no one will add anything without the knowledge of the user; all the information will be shown on a small screen, which will be waterproof and crack resistant.

Inside the bottle will be having sensors to calculate all the materials and minerals present in the water or any fluid or liquid which is to be consumed.

Even it will display if any poison is mixed or any sedatives is mixed, even the dissolved substances will be displayed and shown along with the percentages that will be the kind of accurate measurement and display.

The display and the sensors will work in conjunction and gives the user all the information, quality of water if consumable or not, dissolved quantities and percentages of minerals, metals, solids, etc.

The quality of water will be displayed and can identify the water from government municipal connection, hills and mountains, spring water or processed water or from fresh waterfalls, etc.

Each user can put a pass or pin code to open the water bottle cap for the personal protection and safety at home or office.

The water bottle will be an intelligent water bottle and will display all the contents with whatever liquid it is filled with.

Even a fingerprint can be added to the security features and can locate the bottle if it is lost through the blue tooth feature.

EXPLANATION/BENEFITS/DETAILS/FUNCTIONS

Safe drinking water is essential for life and for everybody because very fewer people can access safe drinking water in the world.

Even in offices, we have seen the bottled water is sometimes not from a reliable source and can lead to sickness and we blindly drink it and to know and be sure that we are consuming safe and healthy water, the check should be there every day and for every fill.

Even the RO processed and UV water can get tested in the water bottle and we can get the results if good for consuming or not.

Kids mostly fill their bottles from the water facility in the school but nobody is sure when was the last time it was cleaned, whether it is harmful or good for the kids, so to be very sure this water bottle helps to consume safe and clean water and maintain their health.

Even the persons going on vacation or picnic and outing can carry the bottle and if the water is available elsewhere and they want to drink, then this water bottle helps in assessing the quality of the water and gives out the information.

We have to go to the depth of each item we are consuming daily, maybe we are risking our lives so to be sure this water bottle helps.

The water bottle will even show the oils and petroleum products and dangerous additives if we want to test, after vigorous testing, only the water bottle will be able to recognise a whole lot of fluids such is the kind of innovation.

Even the bottles will be rented out to the hotels and can be used in offices and homes and schools and hospitals and elsewhere.

Chapter 30

INNOVATION OF SPECIAL GOGGLES

WE CAN SEE THE SIGNALS, FREQUENCIES AND THROUGH THE GOGGLES, WE WILL BE ABLE TO SEE THE CELL PHONE SIGNALS, MOBILE SIGNALS, BLUE TOOTH WAVES, WI-FI SIGNALS, ELECTROMAGNETIC WAVES, VIBRATION WAVES FROM CELL PHONES, SOUND WAVES AND A METER FITTED WHICH WILL DISPLAY THE AMOUNT ACCEPTABLE FOR HUMAN BODY AND BRAIN AND CARRYING WOMEN, ELDERLY PEOPLE AND KIDS.

KIDS PLAYING CONTINUOUSLY IN LAPTOP AND TABLETS ARE SURROUNDED BY THE WAVES, SIGNALS AND FREQUENCIES.

TO ANALYSE IF THE DAMAGE IS BEING DONE, THERE WILL BE A METER OF EXPOSURE OF RADIATION AND TILL WHAT LEVEL IT IS ACCEPTABLE FOR HUMAN BODY.

EVEN THE VIBRATION FROM CELL PHONES CAN BE SEEN THROUGH THE GOGGLES, WAVES SIGNALS OF THE CALLS AND MESSAGES HOW THEY TRAVEL IN SPACE BLUE TOOTH SIGNALS, WI-FI SIGNALS AND HOTSPOT SIGNALS.

INTRODUCTION/PURPOSE

Everyone is surrounded by waves, frequencies, signals, radio waves, vibrations, sound waves, Wi-Fi signals, blue tooth signals, etc. in homes, offices and streets.

The special goggle will help in actually seeing them and enjoying the view.

Even the electromagnetic waves can be seen emitting from household electronics and a meter to measure the allowable range for humans.

EXPLANATION/FEATURES/BENEFITS/DETAILS

We are every time surrounded by infinite signals, waves, frequencies, electromagnetic waves, Wi-Fi signals, blue tooth signals, vibrations and rays, which are not visible to the naked eye.

The innovative goggles will help in identifying and actually being able to see and enjoy the view and to study the pattern of the waves, how they exist, how they behave and how they are received and rejected and their scrambling in the space.

Even the metering and measuring device can show the radiation levels if exceeding the limits of what a human can take.

The goggles will be like that of night vision goggles.

The living creatures can listen to the sounds what humans can't hear with the help of this goggles, we will be able to see the sound waves and analyse.

The waves travelling from a laptop to mobile and PC and radio can be seen.

Even the hearing impaired can see the sound waves and analyse.

Chapter 31

INNOVATION OF BED CAPSULE TYPE

BEDS FOR HOMELESS PEOPLE FOR OCCUPYING AND RELAXING TO PROTECT FROM SUN, RAIN, HEAT AND COLD.

INSTEAD OF SLEEPING ON THE ROAD, THEY CAN RENT AND SPEND THE NIGHT OR DAY TILL THEY HAVE A PLACE TO STAY.

THE TOWABLE OPTION FOR THE BED IS AVAILABLE AND CAN BE TAKEN AWAY FOR SANITATION AND CLEANING.

HELPFUL FOR TOURISTS AND TRAVELLERS; EVEN CAN BE BIG ENOUGH TO ACCOMMODATE ONE SMALL FAMILY.

COMES IN DIFFERENT SIZES, SINGLE AND DOUBLE FOR FAMILY AND BIG FAMILY, RESPECTIVELY.

EVEN THE SMART TV AND RADIO AND WI-FI OPTIONS WILL BE THERE AND IT WILL HAVE PROPER LIGHTING AND VENTILATION AND AC AS PER THE REQUIREMENT.

IT WILL BE SOLAR POWERED AND BATTERY POWERED AND EVEN CAN BE CHARGED BY

THE GOVERNMENTS CONNECTION UNDER THE STREET LIGHT AS DISCUSSED IN THE ROBOTS ROAMING CHARGING POINTS.

SOLAR PANELS WILL BE OPENABLE AND CAN GENERATE POWER GOOD ENOUGH TO GIVE POWER TO THE CAPSULE BED.

INTRODUCTION/PURPOSE

The capsule bed will be an innovation wherein anybody can rent it out, it will be used by the poor people who are homeless and even relatives who are attending a marriage or function or ceremony and doesn't want to hire a lodge and the capsule bed will be towed to their location and parked at the location and the people can spend the night peacefully with all the amenities like fan, AC, light, tv, radio and Wi-Fi signal.

In the morning, they can leave the bed and it will be locked and later can be towed to the place where cleaning and sanitation will take place, allowing it to be able to get ready for use the next day.

The bed is solar-powered/battery-powered with the sponge and a rexine bed and fixed pillows and proper ventilation and heating and cooling mechanisms with all the amenities.

An app will be developed to be able to hire the capsule bed and will be delivered at the location near a park or relatives place or anywhere they feel safe to spend the night and the hire can be for the day as well if needed, they will have the facility to spend the time privately instead of spending on the road.

Reflection stickers and identification number will be seen on it to identify the company and bed and for police checking who have rented and without disturbing the sleepers.

Even it will be floating if the water is suddenly released from somewhere to protect the occupants.

BENEFITS/FEATURES/EXPLANATION/ DETAILS

Homeless people can rent till they have a place to live or even a standard capsule bed can be rented for the accommodation of the relatives where the homes are full in the event of marriages and occasions and ceremonies.

A towable, movable, safe hygienic solar-powered and battery-powered capsule bed will be floatable and even can be paddled to a safe place in the event of a flood.

The main purpose of this innovation is to provide privacy to all and poor who can't afford homes and are spending nights on roads and want to protect from heat, rain, cold, storm, dust and street animals.

It can be towed to the desired location upon the payment and request of the customer and one can spend the night peacefully with a camera for surveillance on the outside.

For most poor who work in the day time and can't afford the house to spend the nights, this is the best place to spend and they then can get back to work in the morning; it will be cheaper than the hotel stay and till they are good enough financially to get them a rent house this is the best place.

Even labours coming from other places to work in the city can rent, put for the night and find a job and till they can afford a place, they can rent out the capsule bed and instead of sleeping in the road and park and benches.

Every year in India, a lot of people die because of heat, cold, rain, storm and lightning strike so to protect those poor people from the natural calamities, this is the best innovation.

The main purpose of this innovation is to provide a safe temporary sleep place to spend the night safely and with the option of displaying the id card on the outside screen and having privacy and respect.

Millions of people especially poor spend the night and day on the roads and they don't have a proper place to protect themselves from heat, rain, storm and cold and risking their lives, at least they get a temporary place till they can afford the rent place for them and their family, even some social work firms can sponsor their stay and even rich can sponsor the stay and support the poor and needy and hardworking labours.

Later in the morning, they can set off to work and the capsule can be towed to a place where regular cleaning and sanitisation will take place and make it ready for the next day slot.

These capsules can be in two versions, one for the poor and one for the visitors who can afford a price and will have a comfortable bed and all the amenities.

When any occasion, marriage or ceremony, is taking place and the relatives are in more numbers visiting the place, then there is a problem of space and this higher version can be rented out and they can spend the night in these capsules and have their privacy.

Even charity can be done by sponsoring the number of capsule beds for a whole month for whoever is in need.

We see poor people are spending on the roads and dying of the weather conditions and nobody is bothered and at least this innovation will give them a place to spend and work in the day time and support themselves even if they don't have a place to stay and still life goes on.

Chapter 32

INNOVATIVE/INFORMATIVE HANDS-FREE SUITCASE

INTRODUCTION/PURPOSE

A suitcase that gives out information and displays weight and displays the ids and itinerary and can be located easily and motorised rolling and following the passenger and having the charging outlets and showing the passports and tickets and visas in the screen and going in sleep mode and even displays the photo of the owner with the help of touch screen.

This suitcase will have a screen and slot for a sim, chip, artificial intelligence and can get connected through the phone to easily locate if lost and can display the banned items when entered the place of travel and tourist places to visit, self-driving wheels and backup power bank for charging the gadgets.

The passenger/owner recognition and following through AI without the effort of pushing or pulling.

The sensors will detect the drugs and will not allow the passenger to carry them to their place of travel and the suitcase will not close and will have an option of fingerprint lock and or pin or passcode to lock it.

The screen on the outside of the suitcase will get the required information and display which is connected through the phone through the blue tooth option like the passport, visa, tickets, ids, etc.

EXPLANATION/DESCRIPTION/FEATURES/BENEFITS/DETAILS

This innovative suitcase will display the exact weight before travelling to the airport and no need to worry about the excess baggage and charges.

Frequent domestic and international travellers are worried about the weight of the baggage, but with this innovation, they will be free of the issues.

The suitcase can be traced with the help of a chip and AI and can be traced from any part of the world through the app and phone.

All the travelling issues will be solved and we can store all the documents in this suitcase and display whenever needed.

When we start packing the suitcase, we just need to enter the place of travel; it will show all the required information through an app and will show the visiting places, tourist places, banned items for that place and best places to shop and best hotels; all the information will be displayed inside the suitcase in a screen and will be helpful.

Even the list of items can be typed which will help in again packing when returning.

The suitcase will have a video recorder and will secretly record if someone is trying to break in as it happens in the case of baggage handlers at the airport.

Even the lost luggage can be traced with the help of the sync facility with the phone, chip and sim.

While collecting at the belt when landed, it will show the location and no need to wait and watch and it will display the pic of the owner as per the setting from the phone.

The video camera can be utilised to take our pics and sending it to the families and it will be a memory of our journeys and travels.

Finally, the innovative suitcase is not a normal suitcase; it's a travel companion with an information display and useful information about the destination and a place to store all the travel documents and power back up soon after landing and self-following motorised and an intelligent masterpiece.

Chapter 33

INNOVATIVE PEN WITH CARTRIDGE

PEN WITH CARTRIDGE SO THAT IT NEVER FINISHES AND LASTS FOREVER

WHEN AN INDIVIDUAL IS USED TO A PEN, THEN IT IMPROVES HIS/HER HANDWRITING BECAUSE OF THE WAY HE/SHE HOLDS IT AND THE HABIT OF USING THE SAME PEN IS IMPORTANT, BECAUSE WHEN THE PEN IS CHANGED, THEN THE HANDWRITING IS AFFECTED.

THE COUNT OF THE NO. OF WORDS WRITTEN WILL ALSO BE DISPLAYED IN THE PEN SCREEN WITH THE RESETTING OPTION USEFUL FOR WRITING EXAMS AND ESSAYS AND THESIS.

INTRODUCTION/PURPOSE

A pen an individual uses should be the same for his/her lifetime because it affects his/her handwriting and the pressure point and he/she gets used to it.

This innovative pen will be of a cartridge type and will give different colours as per the display on the screen and we can adjust the colour setting and options and get the exact same colour on the paper and the words written will be displayed continuously on the screen, even the pencil and the eraser option can be used with the same pen.

BENEFITS/DETAILS/EXPLANATION/ FEATURES

This pen will have thousands of colour options and a pencil and eraser as well and the display of several words and only the cartridge needs to be changed when empty.

This will be mostly used by the kids, students and professionals as there are lots of colour schemes in it and it is a smooth writing one even used by the interior designers to draft their ideas at the quickest and give a sample of their concept design and satisfy the clients and consultants.

Even the format of the words can be selected in the pen if needed, bold, thick or thin.

We can use the pen even in the dark and will have a small led light focusing on the paper or book in which we want to write.

The artificial intelligence concept can be used and whatever we are writing will get stored in the memory of the pen and by synchronising with the tablet or pc, it can make a word document and no need to type in the pc or laptop second time.

Chapter 34

INNOVATIVE SHOES/CHAPPALS WHICH DISPLAY THE NUMBER OF STEPS TAKEN WITH A RESET OPTION AND DISPLAY THE WEIGHT

INTRODUCTION/PURPOSE

Jogging shoes and slippers, home wear will display the no. of steps taken and display the weight in the small screen on the top of the shoe.

This helps the joggers to improve their performance day by day and also the weight displaying will help in monitoring the weight and reducing it.

The AI system in the shoes will check the bp, pulse and display and record it for transferring it to the laptop, pc or gadget, even the date wise weight and no. of steps taken can be recorded and stored for health monitoring.

The shoes will have a capsule for extra comfort and respiration and will take more air and make softer while running and taking heavy steps.

BENEFITS/FEATURES/EXPLANATION/DETAILS

The shoes and slippers will display the number of steps taken in a day or hourly or after every jogging and can be reset; there will be a button on the side to adjust the softness of the shoes as per the requirement along with the memory card and data storing and health monitoring.

These will be water-resistant and waterproof and the capsule will allow to take more air and give extra smoothness and softness.

The daily display of weight and steps taken will encourage the joggers day by day and help them to maintain their health.

The technology used will be a combination of AI, display, data saving, analysing and recording of bp and pulse during peak jogging time and taking the figures of bp and pulse as to how the body reacts.

The led in the shoes will change the colour of the shoes to match the tracksuit and pants.

The shoes will have batteries and be replaceable and a chip and sim are inserted for transferring the data and monitoring the health.

CHAPTER 35

INNOVATIVE TOILET SEAT

AS IN CHAPTER 7, A TOILET SEAT SHOWING WEIGHT BEFORE AND AFTER TAKING A DUMP OR PASSING URINE WITH ADJUSTABLE SEAT FOR DIFFERENT BASE SIZES WITH HEIGHT ADJUSTMENT AND WITH SELF-CLEANING JET SPRAY AND WITH FRESHNER TO CLEAN, NO NEED TO WIPE OR CLEAN USING TISSUE OR WASH WITH HAND.

THE MAIN PURPOSE IS TO IMPROVISE EACH ACTIVITY THROUGH THE 24 HOUR LIFE CYCLE AND MAKE IT MORE ENJOYABLE FOR KIDS, ELDERS AND EVERYONE WHO ARE BORED OF THE ROUTINE LIFE.

RELIEVING AND USING WASHROOM IN OUR HOMES HAS BECOME QUITE A HECTIC PROCESS AND THE IDEA BEHIND THIS INNOVATION IS TO UTILISE THE TIME AND MENTALLY BECOME STABLE.

Chapter 36

LADIES SPECIAL VANITY VAN

AS IN CHAPTER 4, VANITY VAN VISITING NEAR HOME AND GIVING TRAINING FOR LADIES, FOR BEAUTICIANS, PARLOR COURSE AND TAILORING AND ANY SPECIAL TRAINING FOR MORE JOB OPPORTUNITIES; IN THE FUTURE, WE MAY NEED MORE WOMEN DRIVERS AND WOMEN BODYGUARDS AND SO THEY MAY NEED SPECIAL TRAINING LIKE COMMANDO AND FOR SELF-DEFENCE AND PROTECTING SMALL KIDS AND LADIES WHO WILL HIRE THEM AND DRIVING COURSE AND IS USEFUL FOR UNDERQUALIFIED WOMEN.

THE MAIN PURPOSE OF THIS INNOVATIVE IDEA IS TO COME TO THE DOORSTEP EARLY IN THE MORNING AND GIVE SPECIAL TRAINING AND SPECIAL TACTICS FOR WOMEN WHO DOESNT WANT TO TRAVEL BEFORE THE REGULAR JOB AND GET TIRED, THE TRAINERS WILL VISIT THEIR HOMES OR STREET WHERE ALL THE INTERESTED PUPILS CAN JOIN IN FOR THE CLASSES AND BOTH VIDEO AND PRACTICAL CLASSES WILL BE GIVEN ALONG WITH SERVING THE CUSTOMERS AND SEEING IT PRACTICALLY AND LEARN.

CHAPTER 37

LADIES SPECIAL GET TOGETHER IN EVERY AREA INDIVIDUALLY

COMMUNITY GATHERING WITH CATERING AND SOCIAL GATHERING AND MAKING AND IMPROVING RELATIONS AND HAVING A DINNER WITH BUFFET TYPE DISPLAY AND UNITE TOGETHER AND CONDUCTING SPORTS AND MATCHES FOR LADIES ONLY AND SHOPPING MELA AND ITEMS ALL OVER FROM INDIA AND ABROAD AT A CHEAPER PRICE.

THE MAIN CONCEPT BEHIND THIS IS TO MINGLE IN SOCIETY AND GET TO KNOW EACH OTHER IN TODAY'S WORLD WHERE EVERYBODY HAS FORGOT TO INTERACT WITH EACH OTHER AND HAVE A HEALTHY RELATION IN THE COMMUNITIES.

ANYHOW, MEN GO AROUND FOR SOME PURPOSE FOR THEIR REGULAR JOBS AND BUSINESS BUT WITH THE LADIES AND GIRLS ITS TOUGH AND THEY DON'T GET TO GO OUT MUCH OFTEN AND HAVE TO DEPEND ON THE MEN OF THE HOUSEHOLD, SO BY CONDUCTING EVERY MONTH THE GET TOGETHER FOR LADIES ONLY IN EVERY AREA ON THEIR OWN

AND HAVING DINNER AND ALLOW THEM TO MINGLE WITH THE LOCALS AND DO SHOPPING AND A TYPE OF EXHIBITION OF ALL THE CHEAP ITEMS FROM LOCAL AND ABROAD WILL BRING IN SOME TYPE OF REFRESHMENT NEEDED TO RELEASE ALL THE PRESSURE AND FEEL HIGH.

LIKEWISE, IF ALL THE COMMUNITIES GET TOGETHER, THEN THERE WILL BE MORE CONTACTS AND EVEN WOMEN CAN DISCUSS THEIR IDEAS AND MAY BRING OUT THE BEST IN THEM AND FORM A GROUP AND DO SOME SOCIAL SERVICE AND MAY FORM A GROUP AND INNOVATE SOME NEW APP AND DEVELOP AND DO SOMETHING PRODUCTIVE, MAYBE IN THE FUTURE, IF THEY UNITE AND CREATE AN APP, THEN IT WILL BE USEFUL FOR MILLIONS OF WOMEN ACROSS THE WORLD, BECAUSE WE HAVE NEVER ALLOWED OUR WOMEN TO EXPRESS THEIR IDEAS AND MINGLE WITH THE SOCIETY AND NEVER ALLOWED THEM TO DO SOMETHING CREATIVE AND EXPLORE THE WORLD AROUND THEM, INSTEAD WE HAVE KEPT THEM BUSY WITH THE HOUSEHOLD CHORES, MAKING THEIR LIFE BORING AND NEGELECTING THEIR INTELLECT.

GIVING SOME PERSONAL TIME TO DO WHAT THEY WANT ONCE IN A MONTH IS NOT A BIG DEAL, JUST HAVE TO CONTRIBUTE SOME AMOUNT AND ARRANGE FOR THEM THE BEST SHOPPING ITEMS AND BEST CATERING SERVICE AND ALLOW THEM TO ENJOY AND MINGLE AND DISCUSS AND LAUGH AND PARTY AND

ALLOWING THEM TO RELEASE THEIR STRESS IS THE LEAST WE CAN DO.

BY DOING SO, THERE IS UNITY IN THE COMMUNITIES AND LOVE AND AFFECTION AND NO ANTI-SOCIAL ELEMENTS WILL DARE TO CREATE ANY SORT OT ISSUES, WHICH WILL AFFECT THE COMMUNAL HARMONY AND UNITY.

Chapter 38

VENDING MACHINES

APP WHICH SHOWS THE VENDING MACHINES ALL OVER THE CITY

THE MACHINES WILL DISPENSE SANITARY PADS AND MILK PACKS AND HELP THE UNDERPRIVILEGED WOMEN AND KIDS AND CAN BE SPONSORED BY SOME ORGANISATION OR SOMEONE WHO WANTS TO DO CHARITY.

THE VENDING MACHINES CAN BE PLACED AT EVERY STREET AND CAN COVER ALL OF THE CITIES THROUGHOUT THE COUNTRY.

INTRODUCTION/PURPOSE

Vending machines will be placed at every corner and every street of the city and throughout the country; the purpose is to help those underprivileged women who can't afford sanitary napkins and those underprivileged mothers who can't afford nutritious milk for the infants with the help of charity or some social welfare organisation or with the help of some sponsors like celebrities (who are carrying loads of cash and spending on expensive useless attires and apparels and shoes just to look good for the evening party).

The location of the vending machines can be found out through the app and someone can guide the poor underprivileged to the nearest vending machine in emergency night times when kids need milk or women are in need of the sanitary napkins.

It can be placed in front of regular shops and who are willing to help to place the machine and will not charge any rent for that.

The machines will be refilled every morning and evening and will serve the needy.

BENEFITS/PURPOSE/EXPLANATION/ DETAILS

We have seen over the years that the infants and elders may need milk at the peak of the night and can't sleep and even some ladies may have forgotten to purchase the sanitary napkins and may need it in the peak hours, for them the machine will be a lifesaver.

This can be sponsored by anyone and will bring out the humanity in us and the system of the society will be balanced because any society which takes care of itself and its neighbours will never sleep hungry or suffer in need of anything, at the least we can do is to solve the problem of kids and women who are underprivileged.

The vending machines can be sponsored by the big shots monthly and yearly depending on their willingness to help.

Even the machines can be placed at all important public places and help the needy, as we have seen the places where old and used clothes are kept and the poor and needy can pick them up.

Those willing to do charity secretly or openly can sponsor these and can support our fellow Indians and spend their money to support someone's life.

We read in everyday papers that a number of kids die due to non-availability of nutritious food, what a sad and shameful thing to happen in our society where the rich and privileged are worried about the nail polish colour while attending parties, about the matching sandals and about the festivals and shopping, which are few months ahead, and kids, infants, women who are poor and needy and even their tears have dried up are in need of some hope and some support to stay alive at least.

So this innovation will bring out the feeling of at least supporting and the items are not comfort but necessities of life for someone to stay alive, so next time someone sees these machines, they will be thankful for nature as to how they are enjoying life and never wanted to use one of these machines and at least help someone who is in need.

More items can be added up in the future depending upon the demand and can be available in the machine, but as of now, the basic necessities for infants, kids and women are covered.

So the charity options will open up and the machines can be placed on every street and likewise cover the whole nation and at least the death rate of the innocent infants and kids will come down.

The machines will be identifying the regular culprits, defaulters and can be monitored with an inbuilt CCTV camera who are making use of the machines as a business and stealing the items.

CHAPTER 39

NATIONAL PRIVATE TEAMS

ALTERNATE BEST NATIONAL PRIVATE TEAMS IN ALL SPORTS AND SELECTED FROM VARIOUS PARTS OF THE COUNTRY AND MAKING AREA-WISE, STREET-WISE, DISTRICT-WISE, STATE-WISE AND THEN THE BEST OF THE TEAMS IN INDIA IN ALL SPORTS.

BET MATCHES AND PRACTISE MATCHES WITH ANY TEAM WITHIN INDIA AND INTERNATIONALLY JUST TO PROJECT THE TALENT AND TO SHOW HOW MANY STILL TALENTED PLAYERS EXIST.

IF ANY COUNTRY HAVING A POPULATION OF 100 MILLION CAN GIVE BEST TEAMS IN ALL THE SPORTS, WE BEING 1.3 BILLION CAN GIVE ONLY SINGLE TEAMS IN ALL THE SPORTS AT THE NATIONAL AND INTERNATIONAL LEVELS AND YET ARE DEFEATED A LOT OF TIMES.

THE BALANCE IS MISSING.

SO THIS INNOVATIVE IDEA IS TO GIVE THE BEST N NUMBER OF TEAMS IN ALL THE CATEGORIES OF SPORTS AND FOR MEN/WOMEN AND SHOULD HAVE THE FREEDOM

TO PLAY WITH ANY TEAMS NATIONALLY AND INTERNATIONALLY AND THROUGH TALENT SEARCH WILL BE SELECTED AND NOT BIASED IN ANY WAY AND HIRE THEM AND LIKE A REGULAR JOB PAY SALARIES AND GIVE ALL THE FACILITIES AND THEN GIVE VIGOROUS TRAINING AND THEN SEE THE RESULTS.

INTRODUCTION/PURPOSE

Any talented individual can show his/her talent in the sports and when selected will have to work like a regular job and the full responsibility will be taken and given all the facilities and salaries and training and rigorous and vigorous training will make them best.

Any number of players can be there in any sports category and not limited to any number and likewise, from the whole country, hundreds of talented players will come up and cover all the categories, their main purpose will be to be the best in their category of sports but the players will be owned by a private company and a bond will be signed and any time they want to leave and join any other national or state team, then the government will arrange for their transfer by properly compensating the private company owner and then be allowed to break the bond.

State-wise selected people will have to play matches and sports and it will be bet matches and only the toughest will survive, then finally a state team will be made and likewise, all the states will be covered and then the final best out of best teams will be called private Indian team or individual in that category of sport.

State-wise, there will be teams and all the people will be paid salaries monthly and have to play regular matches and the owner will get the amounts and then invest for their developments, training and programs.

BENEFITS/DETAILS/EXPLANATION/PURPOSE

An unbiased talent search across the country and making teams in all sports categories and games for indoor and outdoor.

Those individuals with special talent will have to prove their talent and can get selected for the privately-owned teams and they will get paid and even the pension facility will be available as long as they live.

These people will get all the facilities and even can get their studies completed with the help of the company and their main aim is to outshine their field of specialisation.

The state-wise teams will be the best of the best and can be n number of teams for each state not limited to just one time for each category for each state.

They will be trained accordingly and will be playing matches, sports and games and even finish off their education simultaneously.

In the future, there will be the concept of n number of teams for each sports category and they will be representing the country and all the teams will get the chance to play yearly and the international sports and Olympics and world cups will be held yearly and in each category each time a team is sent and will win all the medals and trophies and cups and we will be the number one in any category of sports in the world.

Or at a time all the teams can go to different countries and can play with the teams of other countries or the foreign teams can come to India and our teams will compete with them in different places simultaneously.

The concept of a single team representing India is not understood and with a population of 1.35 billion, we are not able to give the best teams in the sports and are losing very often at the international level.

The concept is to give 1000 best footballers across the country or give 1000 best batsmen across the country and 100 best wrestlers across the country both in men/women teams, which is possible but not happening because of some reasons and it may happen in the future so just laying the foundation stone of the concept and starting off immediately is the best idea to get the best talents and keep their passion alive and by the time the concept is out, we have our best 1000 in all the categories.

For e.g., if there are 100 categories of sports, then already our 1000 in each category makes 1,00,000 readily available best sportsmen who can't be defeated what so ever and can compete with any international teams.

The companies who want to support and make India a great country should start with the selection process and make private teams and support them and make the selection from even the rural most parts of India in search of real talent and bring out the best and water their talent.

Every now and then, we hear how the talented are left behind and ignored and not given enough chance and lack of amenities and facilities and lack of opportunities make a talent die in one, but the unbiased selection and determination will make real talent grow and when given equal opportunities, it will make wonders.

Chapter 40

INNOVATIVE GOGGLES AND HEADPHONES IN ONE PIECE

TO MAKE GADGET FREE

MOBILE/LAPTOP/COMPUTER/PC INTEGRATED INTO ONE

A WEARABLE GOGGLES AND HEADPHONE WILL COMPENSATE ALL THESE, THE FUTURE TECHNOLOGY LAUNCHING TODAY.

THIS INNOVATION WILL PERFORM ALL THE FUNCTIONS OF ALL THE GADGETS USED BY ANY INDIVIDUAL IN OFFICE AND HOME.

VOICE COMMAND, VOICE-ACTIVATED, VOICE IDENTIFICATION AND VOICE RECOGNITION WILL PRESENT A VIRTUAL SCREEN IN THE FRONT AND CAN PROJECT ON TO ANY SURFACE AND WILL BE TYPABLE, READABLE AND EDITABLE JUST BY A VOICE COMMAND.

SELECTION OF THE SCREENS IN MULTITASKING MODE WILL BE BY THE MOVEMENT OF THE RETINA.

THE KEYPAD WILL BE A VIRTUAL ONE PROJECTED ON TO THE SURFACE OR TABLE OR

DESKTOP AND CAN TYPE JUST BY MOVEMENT OF FINGERS ON THE KEYPAD.

MULTITASKING MULTISCREEN AND FULLY VOICE COMMAND, NO BUTTON OR TOUCH SCREENS LIKE SMARTPHONE.

ALL THE ACTIONS DONE BY ALL THE GADGETS WILL BE PERFORMED JUST BY THE VOICE COMMAND AND THE SELECTION ON TO THE SCREEN IN FRONT OF US WILL BE A VIRTUAL SCREEN AND THE SELECTION CAN BE DONE BY VOICE COMMAND OR BLINKING OR RETINA MOVEMENT, ANY OPTION CAN BE CHOSEN FROM THE OPTIONS.

WE CAN RECEIVE/MAKE VOICE, VIDEO CALLS, CONFERENCE CALL WORK ON THE SYSTEM FOR ANY PROGRAMS, MS OFFICE, BROWSE, OPEN SOCIAL NETWORKING SITES, CHAT IN APPS, ONLINE SHOPPING, TAKING VIDEOS AND PICS, ALL CAN BE DONE AT THE SAME TIME IN THE SCREEN IN FRONT OF US AND IT WILL BE MULTITASKING AND MULTISCREEN.

ALL THE ACTIONS THAT A SMARTPHONE, LAPTOP AND PC CAN DO CAN BE INTEGRATED INTO ONE.

ROAMING HANDS-FREE AND WALKING AND WORKING, GYMING AND WORKING, LAYING AND WORKING, DRIVING AND WORKING, RIDING AND WORKING, SPEAKING TO SOMEONE AND VIEWING THE SCREEN AND LOOKING INTO THEIR FACES AND NO NEED TO CARRY A SMARTPHONE, A LAPTOP, A TABLET, AND

WORRYING ABOUT THE BUYING OF NEW ADVANCED PIECES AND MODELS WHICH ARE RELEASED EVERY NOW AND THEN, WHATEVER THE TECHNOLOGY, IT WILL ADAPT AND UPDATE ITSELF AND THE SCREEN WILL PROJECT IN FRONT OF US IN AIR OR CAN BE PROJECTED TO ANY SURFACE AND CAN BE VIEWED BY ALL, BOTH OPTIONS WILL BE THERE, THE KEYPAD AND KEYBOARD WILL BE AVAILABLE FOR SILENT TYPING WHERE THERE IS SPEAKING RESTRICTION OR SOME CONFIDENTIAL THING NEEDS TO BE SENT.

EVEN THERE WILL BE FACIAL RECOGNITION OF THE PEOPLE AROUND WHICH WILL BE LINKED TO THE DATABASE AVAILABLE WITH THE GOVERNMENT TO IDENTIFY WITHOUT DISCLOSING THEIR NAMES AND OR WE CAN GIVE SOME NAMES AND IDENTITIES TO THE PEOPLE AROUND US LIKE FRIENDS, BOSS, WIFE, BROTHER, SISTER AND TEA BOY WHEN AT HOME, OFFICE OR OUTSIDE.

THE FACIAL RECOGNITION WILL IDENTIFY THE CRIMINALS, LINKED TO THE DATABASE OF CRIMINAL RECORDS, AND SILENTLY ALERT US TO STAY SAFE AND INFORM THE AUTHORITIES.

INTRODUCTION/PURPOSE

By looking at the technological advancements around us, we reached a stage where we can forecast and imagine the technology of the people or evolution of the technology in the next century and innovate those things, which are going to be launched in the future and make use of them today.

In today's world, the sophisticated, accurate, extreme cutting-edge technologies, user friendly, informative and carrying the whole world in our pocket kind of scenario will only work.

So the goggle and headphone will replace all the gadgets, do the same work and be easy to wear and still look stylish.

It will replace all the gadgets in the market and perform all the functions, tasks, multitasking, multiscreen, which will be voice-activated, voice recognition, voice commands, retina scanning for the safety and security features and app and bank transfers and personal data.

The goggle and headphone can be worn and still do the different tasks like driving, riding, looking around, shopping and still keep on working with the visible screen in front of us; when at the office, it will project on to the desktop or screen and the clarity and zoom can be adjustable.

The goggle and headphone will give a screen in front of us, a virtual screen, or will be projected on the plain surface and even while taking a power nap, the screen can be projected on to the ceiling and allows us to work, watch videos and go through the documents and reports of the company, read emails and perform all functions just with voice commands and even the screen can be split into parts for a multitasking feature.

The lens of the goggles will have a camera and the lens will be as per the customer requirement of his sight may be zero or some number of glasses can be used, for taking selfies, a camera will protrude from the headphone and take selfies.

The main idea behind this innovation is wherever we go, all the things should be accessible easily and we should be able to perform all functions and for busy people, this will help a lot and slowly in the future all the gadgets will be compressed into one and no need to buy all the gadgets separately when only one thing can be used and gives us the best performance for whatever task it is being used.

The movement of eyeballs will be read and selection is done on the screen, voice command and voice typing and mail sending without even typing will be possible and in the future, when the chip in the headphone can receive signals from the brain as it is very close, can read our thoughts and can select the task we want to do, it will be a real achievement.

Chapter 41

WINNING OVER THE DEATH

WINNING OVER THE DEATH MEANS THROUGH TECHNOLOGY WE CAN BRING BACK THE DEAD AND MAKE THEM ALIVE IN THE NEAR FUTURE SO BY APPLYING SEVERAL MEANS OF TECHNOLOGY, PRESSURE FILLING OF BLOOD, CREATING EMOTIONAL ENVIRONMENT TO STOP THE SOUL FROM LEAVING THE BODY, WE CAN HEAL THE ALMOST DEAD PEOPLE/BRAIN DEAD AND DELAY THE DEATH.

IN THE NEAR FUTURE, THE SCIENCE AND TECHNOLOGY WILL REACH A STAGE WHERE THEY WILL BE ABLE TO STOP THE SOUL LEAVING THE BODY OR BRING THE DEAD BACK TO LIFE AND DELAY DEATH FOR SOME TIME.

WE HAVE SEEN MEDICAL CONDITIONS WHERE THE PATIENT IS DECLARED DEAD AND MEANWHILE, THE SOUL LEAVES THE BODY AND THEY TRAVEL AROUND AND COME BACK TO THE SAME BODY, SO THE PATIENT WAKES UP, THIS IS BECAUSE NATURE WANTS TO SHOW US ITS POWER.

WE HAVE MANY EXAMPLES WHEREIN THE LIFE EXPECTANCY WILL BE IN WEEKS BUT THE PATIENTS WILL LIVE FOR DECADES.

TO BRING BACK THE DEAD OR HEAL ABOUT TO DIE OR PEOPLE IN COMA ALL THESE CONDITIONS AND SITUATIONS, THERE IS A NEED TO CONNECT WITH THE PATIENT AND CONVINCE THAT HE HAS SOME MORE TIME LEFT THEN THE GUT FEELING WILL OVERPOWER THE DEATH AND THE SOUL RESISTS TO DETACH FROM THE BODY.

AS WE HAVE SEEN MANY MIRACLES HAPPENING WITH THE ORGAN REPLACEMENTS AND ALMOST DEAD AND DECLARED DEAD LIVE FOR YEARS AND NEVER TOUCHED AGAIN BY ANY AILMENTS OR DISEASES OR MEDICAL CONDITIONS.

INTRODUCTION/PURPOSE

Who on earth doesn't want to live a long, healthy and happy life, so the main purpose of discussing the topic is to win over death or delay death or bring back the dead, in any case, this applies to the about to die, dead by some seconds, in a coma for a long time and breathing their last breath and all depends upon the communication with the soul and commanding it to not leave the body and delaying death by some moments/days/weeks.

In the near future, we will see and experience the reality and that day is not so far, for us to communicate with the soul and even scientifically it is proved that A SUPERPOWER/CREATOR is running the universe and has power over everything, so by convincing the superpower or the soul, which is a part of superpower, positive results may be achieved.

To make it easier, when we sleep, the soul travels and is in touch with the body that's the reason we breathe and still dream and react to the movements we are experiencing, but when a person dies, the soul is detached from the body and the body starts to become cold after a few minutes, this few minutes is where the actual thing starts and when the doctors raise their hands, then the SUPERPOWER/CREATOR ONLY has the power to send back the soul into the body and this is the time to convince the SUPERPOWER/CREATOR by means of supplication and asking for more time for the deceased and be vigorous in prayer and make a pledge of any kind helpful for the creation of both mankind and the rest of living things on earth; if it is accepted and the deal looks good, then the SUPERPOWER/CREATOR will send back the soul to the body as he is also called THE MOST BENEFICIENT AND THE MERCIFUL.

Everybody knows miracles do happen but no one wants to put it above all the senses and the people who ask for someone else with the purity of intention, those supplications are heard at once and never rejected.

The main purpose to discuss the issue is every day we see miracles happening and still don't want to endorse the help and give the credit to the SUPERPOWER/CREATOR and soon forget and start boasting and again start our routine and get busy in our work.

When all the doors are closed for help, then one drop of tear from each eye smaller than the head of a housefly, the pain in the heart, the supplication with the purity of intention give us the way out, like help from an unimaginable source and giving more than what we asked for, abundant help from the unexpected source.

Once I read in the diary of a great pious person, "IF YOU HAVE FAITH, HELP WILL COME FROM THE SKIES, THE HEAVENS AND THE EARTH WILL ALL BE IN YOUR FAVOR AND WHEN THE SUFFERING COMES, EVERYTHING GOES AGAINST US," IN BOTH THE CONDITIONS THE SUPERPOWER/ CREATOR WANTS US TO REMEMBER HIM AND ASK FOR HELP LIKE A KID ASKING HIS FATHER FOR MORE CHOCOLATES EVERY TIME, THE POCKET IS EMPTY AND THE FATHER LOVES TO FILL THE POCKET BACK WITH THE CHOCOLATES AND THE TINY HANDS AND THE TINY MOUTH AND THE INNOCENT FACE ASKING FOR FAVOR IS WHAT MAKES EVEN THE CRUEL AND HEARTLESS TO MELT, THE SAME IS THE CASE WITH US AND OUR RELATION WITH THE SUPERPOWER, THE PEOPLE GET ANNOYED IF ASKED BUT THE SUPERPOWER/CREATOR WANTS TO LISTEN TO THE SUPPLICATION AND WANTS TO GIVE THE BEST, BUT THE PROBLEM IS WITH THE CONNECTION AS WE DON'T WANT TO CONNECT TO THE FREQUENCY AND EXPECT THE HELP AND GET ANNOYED AND BLAME AND COMPLAIN ALL THE TIME.

We all are the creation of THE CREATOR and we are the best of the creations with a very beautiful form, shape, colour, appearance, design, structure and arrangement made by the HANDS OF THE CREATOR HIMSELF.

Coming to the earlier point, if a kid asks his father for a break for a day or two after going to school continuously for months, then no father will reject, likewise when the supplication is made with the crying heart and tear-filled eyes, will HE reject who is

THE BENEFICIENT, ALL-COMPASSIONATE/ MOST GRACIOUS/THE MOST MERCIFUL, THE KING/LORD/SOVEREIGN/DOMINION/MASTER/ THE ONE AND ONLY/POSSESSOR OF SUPREME POWER/AUTHORITY, THE HOLY/ALL PURE/ SACRED, THE GIVER OF PEACE, THE GRANTER OF SECURITY, THE ASSURER, THE CONTROLLER, ABSOLUTE AUTHORITY OVER ALL, ETERNAL DOMINATING, THE EXALTED IN MIGHT AND POWER, THE OMNIPOTENT/SUPREME POWER/ POSSESSOR OF ALL POWER/STRONG, THE POSSESSOR OF GREATNESS/SUPREME/JUSTLY PROUD, THE CREATOR/CREATOR OF THE UNIVERSE/MAKER OF ALL/TRUE ORIGINATOR/ ABSOLUTE AUTHOR, THE INITIATOR/EVOLVER/ ETERNAL SPIRIT/WORSHIPPED BY ALL/HAVE ABSOLUTE POWER OVER ALL MATTERS, NATURE AND EVENTS, THE FASHIONER/SHAPER/ DESIGNER/ARTIST, THE REPEATEDLY FORGIVING/ ABSOLUTE FORGIVER/PARDONER/CONDONER, THE ONE WHO IS READY TO PARDON AND FORGIVE, THE SUBDUER/OVERCOMER/ CONQUEROR/ABSOLUTE VANQUISHER, THE ABSOLUTE BESTOWER/THE ONLY GIVER/THE ONLY GRANTOR/THE ONLY GREAT DONOR, THE PROVIDER/SUSTAINER/BESTOWER OF SUSTENANCE/ALL PROVIDER, THE OPENER/ OPENER OF GATES OF PROFITS/RELIEVER/THE VICTORY GIVER, THE KNOWING/ALL KNOWER/ OMNISCIENT/ALL KNOWLEDGEABLE/POSSESSOR OF KNOWING MUCH OF EVERYTHING/ALL

KNOWING, THE RESTRAINER/WITHHOLDER/STRAINTENER/ABSOLUTE SEIZER, THE EXTENDER/EXPANDER/GENEROUS PROVIDER, THE ABASER/HUMILIATOR/DOWNGRADER/POSSESSOR OF GIVING COMFORT, FREE FROM ALL THE PAIN, ANXIETY AND TROUBLES, THE EXALTER/UPGRADER OF ALL RANKS, THE GIVER OF HONOR/BESTOWER OF HONOR/EMPOWERER, THE GIVER OF DISHONOR/THE GIVER OF DISGRACE, THE HEARING/ALL HEARING/HEARER OF INVOCATION, THE ALL-SEEING/ALL SEAR/EVER-CLAIRVOYANT/CLEAR-SIGHTED/CLEAR SEEING, THE JUDGE/ARBITRATOR/ARBITER/ALL-DECREE/POSSESSOR OF ALL AUTHORITY OF ALL DECISIONS AND JUDGEMENT, THE JUST/AUTHORISED AND STRAIGHTFORWARD TO JUDGE OF ALL DEALING JUSTLY, THE GENTLE/BENIGNANT/SUBTLY KIND/ALL-SUBTLE, THE ALL-AWARE/WELL-ACQUAINTED/EVER ADEPT, THE FORBEARING/INDULGENT/OFT FORBEARING/ALL ENDURING, THE MOST GREAT/EVER-MAGNIFICIENT/SUPREME OF/EXALTED/ABSOLUTE DIGNIFIED, THE EVER FORGIVING/OFT FORGIVING, THE GRATEFUL/APPRECIATIVE/MULTIPLIER OF REWARDS, THE SUBLIME/EVER-EXALTED/SUPREME/MOST HIGHEST/MOST LOFTY, THE GREATEST OF ALL GREAT/EVER-GREAT/EVER GRAND/GREATLY ABUNDANT OF EXTENT, CAPACITY AND IMPORTANCE, THE PRESERVER/EVER-PRESERVING/ALL WATCHING/PROTECTOR/GUARDIAN/OFT-CONSERVATOR, THE BRINGER OF JUDGEMENT/EVER RECKONER the one and only who takes account of all matters, THE MAJESTIC/EXALTED/OFT-IMPORTANT/

SPLENDID, THE NOBLE/BOUNTIFUL/GENEROUS/PRECIOUS/HONORED/BENEFACTOR OF ALL, THE WATCHFUL/OBSERVER/EVER WATCHFUL/WATCHER OF EVERYTHING, THE RESPONSIVE/ANSWERER/SUPREME ANSWERER/ACCEPTOR OF INVOCATION, THE VAST/ALL EMBRACING/OMNIPRESENT/BOUDLESS/ALL ENCOMPASSING, THE WISE/EVER-WISE/ENDOWED WITH SOUND JUDGEMENT, THE AFFECTIONATE/EVER-AFFECTIONATE/LOVING ONE/LOVING/THE LOVER/THE ONE WHO TENDERS AND WARM HEARTS, THE ALL GLORIOUS/MAJESTIC/EVER ILLUSTRIOUS/OFT-BRILLIANT IN DIGNITY, ACHIEVEMENTS OR ACTIONS, THE RESURRECTOR/AWAKENER/AROUSER/DISPATCHER, THE WITNESS/TESTIFIER/EVER-WITNESSING, THE TRUTH/REALITY/THE ONLY ONE CERTAINLY SOUND AND GENUINE IN TRUTH, THE TRUSTEE, THE DEPENDABLE, THE ADVOCATE, THE STRONG, THE FIRM, THE STEAD FAST, THE FRIEND/HELPER, THE ALL PRAISEWORTHY, THE ACCOUNTER, THE NUMBERER OF ALL, THE ORIGINATOR, THE PRODUCER, THE INITIATOR, THE RESTORER, THE REINSTATER WHO BRINGS BACK ALL, THE GIVER OF LIFE, THE BRINGER OF DEATH, THE LIVING, THE SUBSISTING, THE INDEPENDENT, THE PERCEIVER, THE FINDER, THE UNFAILING, THE ILLUSTRIOUS, THE MAGNIFICIENT, THE UNIQUE, THE ONE AND ONLY, THE ONE THE INDIVISIBLE, THE ETERNAL, THE ABSOLUTE, THE SELF-SUFFICIENT, THE ALL-POWERFUL/HE WHO IS ABLE TO DO EVERYTHING, THE DETERMINER/THE DOMINANT, THE EXPEDITER/

THE ONE WHO BRINGS FORWARD, THE DELAYER/ THE ONE WHO PUTS FARAWAY, THE FIRST/ THE BEGINNINGLESS, THE LAST/THE ENDLESS, THE MANIFEST/THE EVIDENT/THE OUTER, THE HIDDEN/THE UNMANIFEST/THE INNER, THE PATRON/THE PROTECTING FRIEND/THE FRIENDLY LORD, THE SUPREMELY EXALTED/ THE MOST HIGH, THE GOOD/THE BENEFICIENT, THE EVER-RETURNING/EVER-RELENTING, THE ONLY AVENGER, THE PARDONER/THE EFFACER/ THE FORGIVER, THE KIND/THE PITYING, THE OWNER OF ALL SOVEREIGNTY, THE OWNER/ THE HIGHEST OF MAJESTY AND HONOUR, THE EQUITABLE/THE REQUITER, THE GATHERER/ THE UNIFIER, THE RICH/THE INDEPENDENT, THE ENRICHER/THE EMANCIPATOR, THE PREVENTER/THE WITHHOLDER/THE SHIELDER/ THE DEFENDER, THE DISTTRESSOR/THE HARMER/THE AFFLICTOR, THE PROPITIOUS/ THE BENEFACTOR/THE SOURCE OF GOOD, THE LIGHT, THE GUIDE/THE WAY, THE ORIGINATOR/ THE INCOMPARABLE/THE UNATTAINABLE/THE BEAUTIFUL, THE IMMUTABLE/THE INFINITE/ THE EVERLASTING, THE HEIR/THE INHERITOR OF ALL, THE GUIDE TO THE RIGHT PATH, THE TIMELESS/PATIENT AND FINALLY BY THE NAME ALLAH FOR ME.

I am what I am today and will become what I have to become because of HIM.

This book is in a way the revelation for me from HIM.

My mind was ever peaceful in the most troubled times, most problematic times, the toughest times of my life when there was no possible way out and no other way than the killing of self, I had a vision/trust/faith/belief/connection and listened to the inner voice and had questions and answers as well, and made me a WINNER and inspired to write this book.

I had seen and experienced miracles and this book is not less than a miracle for me, because there was a time I could not speak in front of four people and had stage fear and was shy and sensitive and was always following and wanting to become someone else, but every day for the past three decades, I had witnessed miracles and before that may be my parents had witnessed miracles and help and unknown power over me protecting me all the time and the day has come to pen down my ideas and innovations and feelings and bringing out the best in me and showing others the way and guiding and awakening them and walking together and moving forward and becoming a WINNER.

So finally if you recognise the TRUE POWER and have faith is HIS qualities and have patience, then we can overpower every situation and turn around all the things favourable and will submit to our wish and command, provided if we submit to HIS command and wish and lead a life as per HIS commands and harm no one, hurt no one, like the saying IS THERE BETTER THAN ANYTHING THAN DOING GOOD TO OTHERS, and the best thing to do to others is to show them the right path and help them recognise their values in the eyes of THE CREATOR.

Life and death are the creation and have no power over anybody, they just follow the command of THE CREATOR; when HE is our friend, then how can a true friend harm us?

The main reason is to endorse the reality that human power is to rule the world and THE CREATOR has made everything accessible to us, otherwise, we could not have evolved from carrying stones for making a fire to the pressing of the button to reach moon and mars.

The power of trust, the power of faith, the power of belief and the belief in the power of the SUPER POWER make us unmatchable with any creation in the universe, even the invisible creation exists along with us and are used to coexist and help us without our knowledge, but when the proud, ill-tempered, injustice, suppressing others, killings for power like diseases grow, THE SUPER POWER will destroy the individual/group/community/nation, so to be humble, soft, down to earth are the qualities liked BY HIM, and serving his creation, worrying for others, getting restless for others' problems make us great and good humans and IF HE WILLS what will happen will always be in our favour and HE will make everything accessible and submissive to our command as WE ARE MADE FOR HIM AND THE UNIVERSE IS MADE FOR US, so if we are submissive, the universe will be submissive, otherwise, the same earth will open up and swallow, the same sea will rise and destroy, the water from the skies will drown us, the heat from sun which is to benefit us will burn us, the wind and breeze will create havoc, the pandemics, wars, food shortage, water scarcity, enmity between nations, continuous fear for life, preparing weapons of mass destruction, making nuclear weapons and using them against humanity, dropping bombs on kids and women, killing innocent people, rapes, murders, butchering people, beheading humans for no reason, insulting faiths and religions, stealing from poor, will increase day by day and a time will come we will destroy everything and everybody to get nothing, vanish from the face of the earth and THE SUPER POWER will rise everybody up again to give justice to the oppressed, punish the guilty, give the good their reward, honour the victims, the raped/killed infants, girls, women to become angels and live forever and ever and ever in the never ending universe which is around us and we are like a small drop in the ocean when comparing this earth to the rest of the universe.

Finally, we have time, health, chance to make this world a place worth living and giving the coming generation a better place to live peacefully.

"To kill one man unjustly is like killing all humanity".

CHAPTER 42

BECOMING INVISIBLE OR CAMOUFLAGING CONCEPT AND PROTECTING OUR LOVED ONES/BORDERS/ BUSINESS PLACES/HOMES/ VIPS/HOMELAND

INTRODUCTION/PURPOSE

The requirement to become invisible for the good cause is in demand and to achieve it is the greatest and will be the greatest revolution and achievement for the mankind but it's too late till now; not even anyone is even trying or discussing the concept, whereas it is possible, and with extensive R&D, it will become a reality.

To camouflage with the surrounding and reflecting the same to deceive the vision is the main backbone of this concept, imagine a tree far off and a person is standing in front of the tree and reflecting the same tree in the screen (worn by the person from head to toe and covering him in the shape of a rectangle), we see the tree and think it's a tree and no one will focus on it but the person is actually standing with the same image as of the tree in the screen and confusing the viewer and the person will be able to attack you or take images with an inbuilt camera in the screen and transmit the images and even he can carry arms in case of any danger for his life, even the screen will be both front and back and will cover his entire body and will reflect the same image as present around him in the screen.

Another example, a sofa is in front of you and if a person is camouflaged with the sofa and sitting on the sofa, then we cannot see him or identify his presence and from a distance, it is viewed like a sofa and the person can watch your movements and take your pics and even video record; in this case, if the person is wearing a screen and the sofa is reflected in the screen with utmost clarity as of the original, then it will be more difficult to identify the persons' presence and this concept can be used to protect our borders, VIPs, people, loved ones, business places, expensive showrooms and even help in identifying the rioters who damage public and government properties.

Police and military personnel can be given this special wearable screen and can protect both the country and even avoid riots within the country, because when no one is watching, then lots of unsocial activities take place and start, but actually, by continuous roaming of the protection force, we can avoid any incidents like lynchings, rapes, burglaries, unlawful activities, border crossings of insurgents, attackers from attacking, starting a riot, causing damage to public and government properties, etc.

The protection force can wear the screen and camouflage with the surroundings and never get detected/viewed by the naked eye, imagine a VIP movement in some area where apart from the personal bodyguards need more protection and surveillance if the force can camouflage with the surroundings and stand on roofs or next to trees or stand still without movement anywhere in the vicinity, they can have an eye on the public and can avert any attacks and mishaps and fail the culprits and criminals, even can warn about the movement of the public in the proximity of the VIP.

The main purpose is to make the country safer and protect it from any kind of incidents like mishaps, attacks, mob lynching's, riots, attacks on the police stations, crossing the borders, illegal activities, transporting of arms and ammunition illegally, and on the other hand, it will help in round the clock surveillance without getting noticed and protecting VIPs, shops, banks, our loved ones, valuables, cash transfer to ATMs, jewellery transfer to shops and VIP escort, protecting religious places and turn by turn all the force personnel can take charge in different locations and fulfil their duties and keep the country safe.

Normally, there is a lot of technology in stealth fighting used in fighter planes, stealth bombers, camouflaging effect, or the planes that can fly without able to detect in the radar and complete the task.

The main concept behind this innovation is to mingle in the crowd and yet go unnoticed and keep a check on the activities and any plannings that will affect society.

BENEFITS/DETAILS/EXPLANATION/FEATURES

The concept behind this innovation is that the security/protection teams are in a continuous state of risking their lives when under attack and the intruders will first target them and try to neutralise them and achieve their target.

The wearable screen will be like a protective suit and yet will reflect the images behind it to the viewer on the screen by confusing and camouflaging with the surrounding.

If viewed at night or viewed from far away, no normal eye can detect the force personnel, but the force personnel continues his duty to protect.

The police who will be patrolling by walk in the night can confuse the criminals and intruders and trouble makers by wearing the suit and go unnoticed, they will have an option from inside the suit to record and alert the extra force and will have arms for self-defence in extreme cases.

Even during the night/day patrol for the protection of borders, the intruders normally have the binocular and keep an eye on the security, when absent try to sneak in and cross the border and create a problem by carrying out attacks and killings, which is to be controlled and eliminated totally.

The innovation can be tried, tested and more R&D can be done to achieve the highest value of accuracy in going unnoticed or becoming invisible or camouflaging.

When a force personnel is wearing the suit and moving towards a target and the target notices something is approaching, then the image in the screen will adjust as per the targets' visibility and the clarity improves and will adjust as per the visibility angle and the image appearing in the screen will match the surrounding and behind of the force personnel and transmit with the utmost clarity to go unnoticed and will have a bulletproof vest from inside and a ventilation and comfort air temperature system with juices for drinking and even the urinating without getting noticed.

The concept works in night and day when in dark or in sun, both will have a positive effect in going unnoticed.

The camouflaging concept will help only to an extent, then there will be a real faceoff with the enemy by chasing or running to protect oneself, but the concept of going unnoticed and approaching the target to the nearest point and finishing the target will be a more advanced and safe way.

The bulletproof jacket will help in saving the life, the movement of the force personnel will be on the rollers specially designed shoes to move on hard surfaces and bumpy roads and rocky roads as well, instead of normal walking, to make the movement very smooth, and in normal walking, the movement will be captured by the normal eye and can get attention, a slow and steady movement will never alert the target and can finish off the task and return to the base.

Finally, the wearable suit will have all the features as the location detector, GPS and movement tracker, orientation and can see the target from inside as clear as possible and with the help of a binocular as well, even the person can rotate from inside to go back and forth as per his wish or if found some risk to life, can increase his travel speeds up to 10 miles an hour with the help of roller shoes/motorised power for movement.

The two cameras one in the front and one in the back of the screen will display images on the screens as required, the rear images will show in the front screen and the front image will be shown and displayed on the rear screen to match with the surroundings and go unnoticed and camouflaged.

The screen resolution/clarity will be to the highest precision and with real colours and there will be no considerable amount of difference between the real thing and the image displayed, which will escape from any detection/visibility.

The camera will be of the highest megapixels available and to the highest grade and quality.

If the force personnel is standing still, then no eye can detect it from a distance and ease the job of the personnel, to protect the nation from insiders and outsiders (who create situations harmful to the society) and fail their plans.

The concept can be used even for the protection of VIPs, honourable dignitaries, PM, CMs, MPs, MLAs, MINISTERS, CABINET MEMBERS AND FOREIGN DIGNITARIES visiting.

The secret invisible escort will help in detecting the troublemakers and can easily neutralise them.

The movement will be from a motorised skateboard or roller skates motorised with stability/anti-skid/anti-fall technology.

The screens will be crack-resistant, shatterproof, waterproof and will have a durable life and all the equipment for power and operation will be from inside.

The suit can be used even by the private detectives and security personnel doing their legitimate and legal jobs to keep the surveillance for the companies/warehouses/business places or for the members of the family or to track down the illegal activities and record all the proof and then report to the client.

www.ingramcontent.com/pod-product-compliance
Lightning Source LLC
Chambersburg PA
CBHW020905180526
45163CB00007B/2627